American Colonial Spaces in the Philippines

American Colonial Spaces in the Philippines tells the story of U.S. colonialists who attempted, in the first decades of the twentieth century, to build an enduring American empire in the Philippines through the production of space. From concrete interventions in infrastructure, urban planning, and built environments to more abstract projects of mapping and territorialization, the book traces the efforts of U.S. Insular Government agents to make space for empire in the Philippines through forms of territory, map, landscape, and road, and how these spaces were understood as solutions to problems of colonial rule.

Through the lens of space, the book offers an original history of a highly transformative, but largely misunderstood or forgotten, imperial moment, when the Philippine archipelago, made up of thousands of islands and an ethnically and religiously diverse population of more than seven million, became the unlikely primary setting for U.S. experimentation with formal colonial governance. Telling that story around key figures including Cameron Forbes, Daniel Burnham, Dean Worcester, and William Howard Taft, the book provides distinctive chapters dedicated to spaces of territory (sovereignty), maps (knowledge), landscape (aesthetics), and roads (circulation), suggesting new and integrative historical geographical approaches.

This book will be of interest to students of Cultural, Historical, and Political Geography, American History, American Studies, Philippine Studies, Southeast Asia/Philippines; Asian Studies as well as general readers interested in these areas.

Scott Kirsch is Professor of Geography at the University of North Carolina at Chapel Hill. He is author of *Proving Grounds: Project Plowshare and the Unrealized Dream of Nuclear Earthmoving* and editor, with Colin Flint, *of Reconstructing Conflict: Integrating War and Post-War Geographies* (Routledge).

Routledge Research in Historical Geography

This series offers a forum for original and innovative research, exploring a wide range of topics encompassed by the sub-discipline of historical geography and cognate fields in the humanities and social sciences. Titles within the series adopt a global geographical scope and historical studies of geographical issues that are grounded in detailed inquiries of primary source materials. The series also supports historiographical and theoretical overviews, and edited collections of essays on historical-geographical themes. This series is aimed at upper-level undergraduates, research students, and academics.

Resisting the Rule of Law in Nineteenth-Century Ceylon
Colonialism and the Negotiation of Bureaucratic Boundaries
James S. Duncan

Cold War Cities
Politics, Culture, and Atomic Urbanism, 1945–1965
Edited by Richard Brook, Martin Dodge, and Jonathan Hogg

Micro-geographies of the Western City, c.1750–1900
Edited by Alida Clemente, Dag Lindström, and Jon Stobart

Earth, Cosmos, and Culture
Geographies of Outer Space in Britain, 1900–2020
Oliver Tristan Dunnett

Recalibrating the Quantitative Revolution in Geography
Travels, Networks, Translations
Edited by Ferenc Gyuris, Boris Michel, and Katharina Paulus

American Colonial Spaces in the Philippines
Insular Empire
Scott Kirsch

For more information about this series, please visit: https://www.routledge.com/Routledge-Research-in-Historical-Geography/book-series/RRHGS

American Colonial Spaces in the Philippines

Insular Empire

Scott Kirsch

LONDON AND NEW YORK

First published 2023
by Routledge
4 Park Square, Milton Park, Abingdon, Oxon OX14 4RN

and by Routledge
605 Third Avenue, New York, NY 10158

Routledge is an imprint of the Taylor & Francis Group, an informa business

British Library Cataloguing-in-Publication Data
A catalogue record for this book is available from the British Library

ISBN: 978-0-367-36180-8 (hbk)
ISBN: 978-1-032-43857-3 (pbk)
ISBN: 978-0-429-34435-0 (ebk)

DOI: 10.4324/9780429344350

Typeset in Times New Roman
by KnowledgeWorks Global Ltd.

For my dad, Jeffrey Kirsch

Contents

Acknowledgments

I did not know my (great) Aunt Rose well and did not know her husband—my Filipino uncle, Valeriano Soriano—at all, but their story, fascinating and sad, was almost certainly the first I had ever *heard* of a place called the Philippines. Relations of empire, and the permission of my great-grandmother, an immigrant from Austrian Galicia, had made it possible for Rose and Val to fall in love and marry in New York City in the 1920s. The couple moved to the Philippines in the early 1930s and Val worked in what must have been an established family business, for he also served, from 1934 to 1937, as mayor of the municipality of Orion in Bataan province, where he was known, according to a 1950 town plaza dedication, as American Boy Soriano. Rose (perhaps also Val) survived the Pacific War in what she described as a brutal Japanese internment camp, but their life together in the new Philippine Republic would be cut short when Val was assassinated, alongside Francisco C. Quezon, in an execution-style killing during an unsuccessful raid by *Huk* guerillas in 1948. These events occurred well after the initial American colonial period that I attempt to reconstruct in this book, but they were in many ways made possible by the earlier imperial moments; they offer a sense of the historical forces at play in U.S.–Philippine relations throughout the twentieth century as well as the planetary dimensions of human experience, and no doubt they sparked my interest. Rose Thall Soriano, for her part, lived into her nineties, until her later years splitting time between New York and Luzon.

Research for *American Colonial Spaces in the Philippines* was funded initially under a National Science Foundation grant (BCS-0518213) supporting archival research at the National Archives in College Park, Maryland; Library of Congress (Geography and Map Division); University of Michigan's Special Collections Research Center and Bentley Historical Library; Ryerson and Burnham Archives, Art Institute of Chicago; National Library of the Philippines; and American Historical Collection, Rizal Library, Ateneo de Manila University. Additional support from the Newberry Library (Chicago) and Houghton Library, Harvard University, allowed me to spend time with their wonderful collections. The work of archivists and librarians at all these sites did much to make "American colonial spaces" visible to me. I am also grateful for institutional support from the University of North Carolina's College of Arts and Sciences and Institute for the Arts and Humanities, for the supportive home that the Department of Geography has provided for teaching and research, and for colleagues in Chapel Hill and places beyond whose friendship and intellectual engagement has meant a lot to me throughout a lengthy project, but who will go unnamed.

Special thanks are due to three former doctoral students who, in the course of moving on to bigger and better things, made unique contributions to the project. Craig Dalton's database of my NSF research, constructed over a decade ago, was indispensable, particularly after 2019 when I wrote most of the book. In addition to research assistance in the Philippines, Joseph Palis hosted my visits to the Department of Geography at the University of the Philippines Diliman in 2008 and 2015, introducing me to the Philippines of the present even while helping me chase the ghosts of American colonial spaces. I am also grateful for the warm reception and feedback from UP colleagues and students during talks in the department, with special thanks to Lourdes Benipayo for kindly showing me the remains of Baguio's American landscapes, and to Jun Aguilar of Ateneo de Manila University for his hospitality and interest in my research. Finally, Mike Hawkins, while writing his own dissertation, an "episodic historical geography" of the Port of Manila from 1898 to 2020, did me the huge favor of reading the entire manuscript in draft, and I am grateful for his generous reading, suggestions, and insights.

Parts of this work grew from earlier publications which have been reworked and expanded upon in this book. I appreciate the kind permission of journal editors to make use of these articles from *Philippine Studies: Historical and Ethnographic Viewpoints* ("Aesthetic regime change: The Burnham Plans and U.S. landscape imperialism in the Philippines" from Vol. 65 (2017), pp. 315–356) and *The Geographical Journal* ("Insular territories: U.S. colonial science, geopolitics, and the (re)mapping of the Philippines" from Vol. 182 (2016), pp. 2–14), as well as their input and that of their reviewers in the earlier work. Parts of a book chapter from the collection *Reconstructing Conflict: Integrating War and Post-War Geographies*, eds. S. Kirsch and C. Flint ("Object Lessons: War and American Democracy in the Philippines," Routledge (2011), pp. 203–225), are also included in revised form in the present work by permission of the publisher. I am also thankful for Faye Leerink, Prachi Priyanka, and the team at Routledge for making the process so easy.

It is my pleasure to dedicate the book to my dad, Jeffrey Kirsch, with love, admiration, and gratitude. An aerospace engineer from Brooklyn turned public television producer and host, science museum director and film producer in San Diego, and purportedly Aunt Rose's favorite nephew, he encouraged me to work hard and to follow my curiosity, giving me the confidence to do so along with his own model of how that might be accomplished.

Illustrations

Figures

Table

Introduction

An Imperial Banquet

On a Tuesday evening in May 1912, the captains of the American railroad and banking industries gathered at Sherry's restaurant in Manhattan, a posh Fifth Avenue banquet hall, with a ballroom of ruling class scions that included three Tafts and two Vanderbilts. They had come to fete the return of one of their own, the Governor-General of the Philippines, W. Cameron Forbes. Following hors d'oeuvres *a la russe*, a soup of *boeuf anglaise* preceded the sole *belle helene*, spring saddle of lamb with lima beans and parsleyed potatoes, Easter ham and salad with fine herbs, asparagus with hollandaise, and if any had saved room, a dessert of Sherry's famous *glace fantaisie* with coffee and cakes. The purpose of the evening, apart from, evidently, massive caloric intake among a crowd of like-minded, raced, dressed, and gendered bodies (Figure 0.1), was to *meet* Forbes, as the invitation from a dozen influential New Yorkers had put it simply, during the Governor-General's stateside visit. The program that followed, including speeches from New York City Mayor William Jay Gaynor and former Senator Chauncey Depew, coupled a celebration of perceived American accomplishments in the archipelago with, from Forbes himself, an insider's update on "conditions in the Islands," a 30-minute speech styled as "plain talk," spoken "man to man, like a businessman," part sales pitch and part geopolitical argument, promoting investment in the Philippines—and the entrenchment of U.S. strategic and economic interests—for the foreseeable future.[1]

In some ways, the speech resembled countless others that Forbes had delivered since his appointment to the Insular Government (as the civilian branch of the U.S. colonial government of the Philippines was called) in 1904, at age 34, as Secretary of Commerce and Police with a seat on the Philippine Commission, and as Governor-General from 1909. Forbes was a prolific, if, by his own measure, less than scintillating, speaker. He spoke at the Army-Navy Club and Manila Hotel on the Daniel Burnham-planned bayfront esplanade in Manila and at the City Club of Boston and Lake Mohonk peace conferences during North American visits. He spoke at banquets held in his honor during endless "inspection" tours of the Philippine

DOI: 10.4324/9780429344350-1

Figure 0.1 Dinner in Honor of the Honorable W. Cameron Forbes, Governor-General of the Philippines, Sherry's, 14 May 1912.

Source: Courtesy, Houghton Library (MS 1365.1), Harvard University.

provinces, a duty which Forbes believed should properly occupy one-third of the Governor-General's time. Whatever the audience, Forbes was an unyielding apologist for the "strictly self-supporting" American-led Insular Government who found it difficult to find fault with the American imperial project under any circumstances, though he felt obliged at times to complain of inefficiency or mismanagement, particularly among Filipinos, in government. At Sherry's, Forbes touted opportunities in the Philippines for Americans with talent, initiative, and grit; praised the quality and value of Filipino workers when paid a living wage; and commended the Filipino people on their progress under American tutelage. Avoiding the language of 'race' and 'capacity' that underwrote much U.S. colonial policy, Forbes expressed confidence that the American venture in the Philippines would ultimately be profitable for everyone.[2]

As he explained the relative *lack* of progress in Philippine agriculture, manufacturing, and commerce compared to rosy American projections of a decade prior, the main problem was one of circulation, for "at almost every point in the flow of products from the soil to the consumer there were placed obstructions of such a nature that the value was checked until finally only the littlest bit of trickle reached through." How else to explain that Hawaii produced 30 times the export earnings of the Philippines, Forbes asked, and "Porto Rico" six times the Philippines' export value? Forbes was not wrong that exports had been held in check in the archipelago which, until the Payne Act in 1909, did not enjoy the "normal" benefits of free trade with the United States, a decade after Americans claimed sovereignty over the archipelago by conquest and treaty (and a sum of 20 million dollars) with the Kingdom of Spain. As Senator Depew, a former Vanderbilt railroad lawyer, would point out, this condition of blockage, wrought by illiberal trade practices, had been foisted on Filipinos, reflecting the power of protectionist U.S. sugar interests, even as planters' once principal market (Spain) had been cut off from them. Now, emphasizing material improvements in the conditions of exchange through construction and maintenance of roads, bridges, wharves, and harbors, alongside the establishment of classical colonial trade relations, Forbes offered a vision of enduring profitability through the re-orientation of the Philippine resource economy around exports of sugar, rubber, hardwoods, chocolate, hemp, tobacco, copra, and pearls: "All we need is an improvement in capital."[3]

Claiming to have achieved a "perfect public order" under the Insular Government, the onetime Harvard football coach Forbes justified a recent U.S. Congressional request for additional road construction financing, by urging, "We can build the country faster." Whether speaking as "we" the Insular Government, the assembled party at Sherry's, the Filipino people, or all of the above, Forbes expressed a profoundly ideological sense of historical agency. Who was building the country, and under what conditions? What were they building the country for? Forbes mapped the Philippines as an "asset" linking geopolitical and geo-economic dimensions of American empire just as the Pacific was emerging as the "great theater of the world's commerce." Evoking the Philippines' proximity to the "sleeping giant" across the South China Sea, Forbes asked his audience to "Think of what is to the US, to have that place at the Gateway to China," imagining a landscape of "American warehouses in Manila, stored with American goods," that would allow "us to reach and get our share of the Chinese trade but ready for the Chinese development."[4] Whether *having* that place meant it was a formal part of the United States or "a friendly or allied power … tied only by commercial ties," Forbes submitted, "I don't pretend to say," but he added that he saw no special virtue in "pulling up the flag" and "no advantage to giving up the place we have won with so much treasure and blood, advantageous to us and them."[5]

While the ascendancy of a planetary American empire, as characterized by geographer Neil Smith to be ruled principally by markets rather than direct territorial control, was delayed until the aftermath of World War II, emerging and conflicting forms of U.S. imperialism took shape in the "Insular Territories" of the Caribbean and Pacific around an earlier moment of U.S. global ambition.[6] The Philippine archipelago, made up of thousands of islands and a diverse population of more than seven million people, was the unlikely, primary setting for experimentation with formal colonial governance after the Spanish-American War, though it would require years of guerilla warfare, counter-insurgency, and military occupation (until 1913 in the southern archipelago) for the United States to maintain its sovereignty there, and even then for a relatively short time. *American Colonial Spaces in the Philippines* tells the story of a regime of U.S. colonial retentionists that attempted, in the long first decade of the twentieth century, to build an enduring American empire in the Philippines through the production of space. From concrete interventions in infrastructure, urban planning, and built environments to more abstract projects of mapping and territorialization, the book traces the efforts of U.S. Insular Government agents to transform the Philippine landscape, alternately brutal and aestheticized, imperialistic and democratic, practical, fanciful, and draconian. In doing so, it also attempts to shed light on the limits to these spatial strategies, and their appropriation by different actors for different ends, raising questions about both the precariousness and the persistence of America's twentieth century insular empire.

Few of the limits to empire were on view at the spring affair at Sherry's, a venue well-known enough that its address—at the corner of Fifth Avenue and 44th Street—was not included in the invitation. Transcripts of the evening's remarks reflect instead a cheerful levity among the 150 "leading men" and "representative New Yorkers," as the *New York Times* described them in the next day's paper, who had gathered to celebrate Forbes, and to offer their tacit acceptance of America *as* empire, engaged from a sated and comfortable distance.[7] Rather than Forbes's "detailed picture of the remarkable progress" achieved during his eight years in the Philippines, however, it was Gaynor, the Democratic mayor who had already captured public attention by resisting control from Tammany Hall—and surviving a gunshot to the neck—who merited the lede for *Times* headline writers when he predicted the "most terrible conflict ever known" on the horizon with China in his after dinner comments.[8] Reflecting widespread views of impending, inter-civilizational and racial conflict in the Pacific, Gaynor insisted in advance that, unlike the European states, "our hands are clean," before turning his attention to the honoree Forbes, who had perhaps provoked the outburst with his discussion of sleeping giants. The transcripts suggest in Gaynor an orator in command of the room—parenthetical "laughter" is noted seven times in the short speech—wondering aloud how best to honor the returning Governor General:

> When I see this young man here tonight, with his fine profile … this Governor from the Eastern Hemisphere. I don't know, Mr. Forbes, whether to call you a Proconsul or a Procurator [laughter], but inasmuch as such extraordinary powers are vested in our President, probably equal to—no, certainly equal to and maybe greater than those vested in the Caesars, I think I will salute you as Procurator come home here to visit us tonight, and so I have looked upon this young Pontius Pilate.[9]

Another roar of laughter, amid which Gaynor is heard to confess that it was the only Roman procurator he could think of at the moment, clarifying that Forbes was of course someone "in the way of helping that part of the world materially," but it was too late; a running joke had been born. Speaking later of diplomatic endeavors in Europe, Depew joins in the fun, referring to a "Pontius Pilate view of public duty," to which "a voice" from the audience calls out, "Pontius Pilate wasn't such a bad fellow at that." Yes, Depew responds, seizing on the tagline to a new round of applause, "Pontius Pilate was exactly like the Governor-General—he tried to help the people."[10]

Forbes makes no special notice of the jests in his sometimes expansive, sometimes prickly, personal journals, in which he describes the evening only as "a dinner offered by a group of capitalists and business men" at which he "talked for an half hour on Philippine affairs to a seemingly appreciative audience… representing the largest railroad and banking businesses and it was certainly a bunch of men worth meeting."[11] Neither the hubris of the moment, nor Gaynor's choice of Forbes as procurator of an *empire*, rather than proconsul serving a Republic, escaped the notice of the nationalist press in Manila, however, when details of the evening reached the Philippines. In an editorial in *La Vanguardia*, the newspaper credited Gaynor, on the "occasion of an imperial banquet given in the US commercial metropole," with either a "sense of genius" or "moment of clairvoyance" in dubbing Forbes Pontius Pilate, allowing that it must have felt good for the celebrants for a few moments to enjoy the "blue and glorious blood of the Roman citizens" coursing through their veins. The parallels between the ancient Roman and modern American empires were not difficult to establish, the anonymous article argued; more surprising was the frank tone of Forbes's remarks, which lacked the "altruistic feelings toward the Filipino people" conventionally expressed in his Philippine speeches. His comments at the banquet offered instead a clear picture of how the archipelago was seen by Americans: as "an advantageous point" from which to reach out to obtain greater wealth and opportunities. But if ancient Rome belonged to the past, "mined by its vices, victim of its disordered ambitions," the author cautioned, "the modern Rome, with its Pontius Pilates who know how to properly wash their hands, remains imperturbable in its careless career toward glory and power."[12] Observing that Senator Depew appeared so delighted with the nickname that there seemed to have "never been a mayor so witty as Mr. Gaynor of New York," *La Vanguardia* reported in closing that "the chronicle does not say if Mr. Depew asked Mr. Gaynor for the health of Christ crucified by order of Pontius Pilate."[13]

It was hardly the first time that Forbes had been skewered in the pages of the Manila press. Within a year of his inauguration the *Free Press* had already questioned—in both text and cartoon—whether "the Forbes' sun" was rising or setting.[14] More recently, *La Vanguardia* had offered a scathing "theory of intussusception" that likened the Insular Government's special treatment of a cast of "Americans and foreigners" in Mindanao to a disease of intestinal blockage, invoking the medical term describing the inversion of one portion of the intestine within another.[15] The possibility of Forbes not returning to the Philippines from his Stateside vacation had been a matter of open speculation even before his departure from Manila, and was given fresh life after a run of health concerns compelled Forbes to extend his leave for an extended period of convalescence. While President Taft warmly urged Forbes in July to take all the time he needed to recover his health and restore his "early enthusiasm" before returning to the Philippines, it was clear, as *La Vanguardia* had surmised, that Taft was in the fight for his own political life, and that the political fortunes of "the young Pontius Pilate, that is, Mr. Forbes," would depend on those of "Papa Taft … in the November elections."[16] The dinner at Sherry's provides a window onto the imperial moment as celebrants, even the reluctant colonialists plied at last with coffee and cakes, enjoyed the pleasures of their empire of plain talk, looking ahead to a future in which success seemed assured.

The dream of a lasting American colonial empire in the Philippines, whether cultivated in after dinner speeches or built into brick-and-mortar landscapes, would be a difficult one to hold in place, and real historical change was scarcely anticipated. The horizons of individual careers could be equally difficult to envision. Within six months of the banquet at Sherry's, Taft had been roundly defeated by Woodrow Wilson, as ex-President Roosevelt's Bull Moose ticket split the Republican vote, and Forbes's career in the Philippines, and the nature of U.S. colonialism under the ostensibly anti-imperialist Democratic party, would be called into question. Forbes, meanwhile, would struggle to regain his health after being diagnosed by physicians with a deterioration of muscles around the heart, stemming from an infection the previous year, and a planned recuperative visit to the high country in Wyoming had only made matters worse.[17] William Jay Gaynor would die suddenly on board a steamer to Europe in September 1913, perhaps the victim of an unretrieved piece of the assassin's bullet that had been lodged in his neck for the intervening three years. Although the inter-civilizational conflict with China that the New York mayor predicted failed to materialize, his view of the emerging American empire, with imperial powers, vested in the President and War Department, exceeding those of the Caesars even while "washing its hands" of the imperial taint, though morally suspect, was not wrong, and could be glimpsed taking shape around a liberal "empire of no empire" even within formal U.S. colonial projects.

American Colonial Spaces in the Philippines presents a multifaceted historical geography of American empire and colonial life in the Philippines, focusing on the period from 1900 to 1913 when a governing Taft-Forbes regime, aligned closely with the U.S. Republican Party, occupied the upper echelons of government and instituted key elements of these spatial policies. Drawing on and elaborating notions of the production of space associated with the Marxist philosopher Henri Lefebvre,[18] the book traces the production of four *kinds* of colonial space—territory, map, landscape, and road—that were each geared, in different ways, to make space for American empire in in the Philippines and, some hoped, to help ensure its survival for a few generations at least. In the next section, I address more explicitly what it means to "produce" space in this context and begin to tease out the relations among the production of space, 'state space,' and ideologies of U.S. imperialism that were bound together in efforts to reproduce the Philippines as a distinctively *American* colonial space. While the elite U.S. colonials who sought to fashion these spaces were by no means acting alone—consider the everyday labor required to build and maintain roads and culverts, or the efforts of elite Filipino families, on whose support Insular officials relied, to co-opt U.S. policies as a means of entrenching their own local and national power—the Americans remain, for practical and analytical reasons, the chief protagonists of my story.[19] But while the experiences of key actors such as Forbes are central to its narrative, the book also relies on a "regime approach" to political agency, hence it understands political power in colonial administration as relationally distributed in an evolving but more or less stable coalition of actors and institutions, based in Manila and Washington. Meanwhile Taft climbed the ladder of positions in charge of U.S. Philippine policy from Governor-General to U.S. Secretary of War to the presidency, with Forbes ultimately rising to the Governor-Generalship. Tracking this regime's efforts to produce spaces of sovereignty (territory), knowledge (maps), aesthetics (landscape), and circulation (roads) in the Philippines, as part of an evolving complex of colonial, geopolitical, and class relations, provides the key conceptual structure for the book.

Taft-Forbes Regime and the Production of Space

> What is an ideology without a space to which it refers, a space which it describes, whose vocabulary and links it makes use of, and whose code it embodies? … More generally, what we call ideology only achieves consistency by intervening in social space and its production, and by thus taking on body therein. Ideology per se might well be said to consist primarily in a discourse upon social space.[20]

American Colonial Spaces does not attend closely to the complex social and cultural relations of the Spanish period, extending over hundreds of years of territorially uneven colonial rule, through which the very idea of *Las Islas Filipinas*, as a distinct political unit, was produced, and enters my story more or less fully formed.[21] What is more, the principal approach adopted—following elite actors from the U.S. regime whose records, maintained within an emerging imperial bureaucracy and through personal archival practices, have been well-preserved—risks replicating or over-emphasizing those perspectives, their categories of knowledge, and their sometimes grandiose sense of self-importance. Hence, in addition to limits of historical breadth, epistemology, and expertise, the book is clearly an outsider's perspective on the Philippines, and with its peculiar focus on things American, an insider's view of U.S. empire. In this respect, it is clear to me that the narrative that follows is not entirely unlike the perspectives on view at the Forbes dinner at Sherry's, even if intended as critical rather than celebratory in form.[22] Despite these drawbacks, I suggest that an open and situated "production of space" framework, built around the critique of ideology and social space, offers some compensation for addressing these limitations, and an effective means of interpreting the construction of American colonial spaces in the Philippines as complex and unstable achievements. The approach sheds light on the making of "American" colonial spaces—parsed as spaces of sovereignty, knowledge, aesthetics, and circulation—that reflected and reinforced existing power inequalities but also generated new contradictions of empire, and conditions for change, at multiple sites and scales.

Cameron Forbes was more than an apologist for the Insular Government in the Philippines; he was among its most motivated agents. For Forbes, who supervised the Insular Bureau of Public Works during his five years as secretary of Commerce and Police and subsequent four years as Governor-General, and pushed an aggressive roadbuilding and maintenance agenda, the practical and symbolic dimensions of (re)building the Philippine landscape were closely entwined. As Forbes summed up his Philippine career to that point for President Taft, on the eve of his inauguration as Governor-General in 1909, "I have made material progress my slogan throughout the Islands ever since I first arrived here." As a result, Forbes crowed, "I am greeted as the apostle of public works and material advancement," with the added bonus that "in my presence, people very seldom refer to political matters."[23] Who could deny that a gravel road where there had been a dirt trail, or bridges fashioned with reinforced concrete and steel, constituted progress under American leadership?

As historian Paul Kramer observes, notions of "building," "construction," and "constructiveness" were mainstays in the vocabulary of the American colonial cadre in the Philippines during the first decade of U.S. rule, and for Forbes in particular.[24] The terms served as powerful metaphors of colonial nation building, while also signaling key policy priorities that crystallized around large-scale infrastructure projects and the distribution of state architectures across the archipelagic landscape. These projects offered evidence, for advocates, of benevolent and abundant U.S.-led modernization programs in the Philippines, carried out under a broad rubric of "material development." State interventions in landscape, while invested in the signature landscapes of Manila and Baguio, the American "summer capital" in the highlands of northern Luzon, and later in provincial capitals of Zamboanga and Cebu, were also widely distributed in schools, municipal buildings, and public markets throughout the archipelago, rendering new hierarchies out of the very materials from which public buildings were fashioned and producing new surfaces of interface between people and the colonial state.[25] The regime's preference for reinforced concrete in its construction projects attached a visual sense of permanence to many of its buildings, Kramer argues, such as the schoolhouses that in provincial regions provided "the most intensive marker of the presence of the U.S. colonial state." Hence, for Kramer, "Despite Filipino funding and labor committed to construction of schools, these structures and others were meant to contrast sharply with 'native' buildings. This nation was being constructed from the top down, by a state that stood architecturally apart from its subjects."[26]

The widely used phrasing of the "strictly self-supporting" Philippine state, so defined on the basis of revenue collection that funded its operations, turned on a similar ideological equation, though its ledgers could only be squared by externalizing various military, security, and transportation costs, and further excluding the costs of the bloody Filipino-American war, from empire's moral arithmetic. As Forbes would fire back at critics of the Insular Government's expenditures in his May 1912 comments at Sherry's, "People in the United States talk of the great cost of the Philippine Islands, they speak of how expensive they are, but they seem to have confused the actual cost, the past payment, for that cost is now over. The big expense is past. It is something like making all the charges on the debit side without adding any entries on the credit side, without giving the account a chance to balance up."[27] In this coding of geopolitical sunken costs, the "strictly self-supporting" Insular state relied, like the notion of a distinctively American colonial landscape, on ideological boundary-work, achieving consistency, as Lefebvre argues, precisely "by intervening in social space and its production."[28] What did it mean then to make space for American empire in the Philippines?

Lefebvre's writings on space were extensive and heterodox, but for our purposes can be summarized briefly. At his most suggestive, Lefebvre asked how the production of space made possible the survival of capitalism over time, and in subsequent writings he outlined the pivotal role of the state in these processes.[29] This book extends a related question, hinging on this "reproductionist" reading of Lefebvre, in the context of U.S. interventions in the Philippines: how spaces and landscapes were produced to enable the survival of a formal American empire amid a range of forces, from Filipino revolutionaries and nationalists to American protectionists and anti-imperialists and the looming Japanese empire, that all threatened to undermine it (and eventually did). Reading key forms of territory, maps, roads, and landscapes "through" the production of space, in this sense, with distinctive chapters dedicated to each spatial category, can help to illustrate both the transformative capacities of imperialism and the instability of spatial strategies of empire, and the survival of empire and state space in new forms.

While the book is intended, through this framework, to contribute to broad conceptual questions, its main emphasis is substantively on *these* spaces in the early twentieth-century Philippines, and on the efforts of the Taft-Forbes regime to demonstrably remake them. It is derived from the colonial archive (among other sources), constituting a narrative of real people, places, and events that is meant to help rethink the American imperial moment in the Philippines with an explicitly spatial vocabulary, tracking maps, plans, and projects of a relatively small regime of fin-de-siècle U.S. imperialists. Given this "top-down" approach, a reconsideration of the concept of ideology from Marx's (and Lefebvre's) critical toolbox offers not just a useful reminder to avoid uncritically reproducing the world-views of our key informants, though that is obviously important, but also raises questions about how the erstwhile science of ideology can help us to rethink the production of space in terms of colonial power relations in specific historical-geographical contexts.

Keyword: Ideology

Although ideology (and the critique of ideology) are fundamental in his arguments about the production of space, as well as earlier efforts to imagine a Marxian sociology of knowledge, Lefebvre expressed an ambivalent relationship with the term. If the "concept of ideology" was "one of the most original and most comprehensive concepts Marx introduced," offering new ways of relating social representations to the worlds in which they were produced, then it was also one of his most complex and obscure.[30] In *The Production of Space*, Lefebvre would find "the long obsolescent notion" of ideology to be "now truly on its last legs, even if critical theory still holds it to be necessary."[31] But this apparent paradox was resolved, for Lefebvre, in space. By opening the material, symbolic, and experiential dimensions of space to questions of ideology, that is, and integrating an effective critique of ideology into the broader analysis of the production of space, Lefebvre offered an integrative focus on the *work that space does* in regimes of domination and power, whether ruled by capitalist, colonial, or nationalist logics, or combinations thereof.

Lefebvre traces "ideology" to the *idéologues* of late eighteenth and early nineteenth century France, a philosophical school ("empiricist and sensationalist, with a tendency to materialism") that sought to understand the basis of ideas or concepts in bodily sensations.[32] Yet Marx, working with Engels, "transformed the meaning of the term," or perhaps endorsed a transformation already underway, in which the term "now became a pejorative one. Instead of denoting a theory, it came to denote a phenomenon the theory accounted for. This phenomenon now became a collection of representations characteristic of a given epoch and society." Some of the original meaning of ideology—the erstwhile science of ideas—inhered, for Marx still "aimed at formulating a theory of the general, i.e., social representations; he defined the elements of an explanatory genesis of "ideologies" and related the latter to their historical and sociological conditions."[33] In this critical turn, once the realms of representation—in philosophy, religion, art, and "knowledge itself"—could be understood as relative to situated, earthly conditions, a closer look at the different kinds of work that representations do in the world became possible. In Marx's treatment, ideology, expressed as the *critique* of ideology, is reinvented as a "theory that generalizes special interests—class interests—by such means as abstraction, incomplete or distorted representations, appeals to fetishism." For Neil Smith, building on Lefebvre's definition in the context of his related production of nature thesis, ideology was thus "not simply a set of wrong ideas but a set of ideas rooted in practical experience, albeit the practical experience of a given social class which sees reality from its own perspective, and therefore only in part. Although in this way a partial reflection of reality, the class attempts to universalize its own perception of the world."[34] But while the Marxian concept of ideology offered a productive framework for interpreting the emergence and distribution of ideas that served the interests of dominant groups or were acceptable to them, it did not, for Lefebvre, go far enough in elucidating how this generalization of special interests took shape, or what was at stake in this process.

Lefebvre's reworking of ideology was centered on the production of space, broadly conceived to include material and symbolic dimensions of what cultural geographers commonly call place and built environment. On the terrain of religion, the analogy was precise:

> What would remain of a religious ideology … if it were not based on places and their names: church, confessional, altar, sanctuary, tabernacle? What would remain of the Church if there were no churches? The Christian ideology … has created the spaces which guarantee that it endures.[35]

Extending into the underworld through subterranean crypts, and pointing with their spires toward heavenly perfection, the built environments of Christian sacred space thus offered lessons on the everyday work of ideology, understood in terms of a universalization of perspective, or generalization of special interests, but also as a particular "discourse upon social space."[36] Ideology did not belong to a separate, immaterial world of ideas (as if such a thing existed) but rather reflected a powerful coding of built environments in the "concrete abstractions" of social space.[37] His point is not only that the Church produced its own functioning, symbolic world, but to ask what it produced that world *for*. Hence, for Lefebvre, the point of the production of space concept was not merely that societies transform—and in a sense produce—their own dimensions of space, place, and environment, but rather that capitalism had persisted only by ceaselessly producing the space(s) of its own survival. And social space, at once the outcome of past actions and "what permits fresh actions to occur, while suggesting others and prohibiting yet others," depended on a great deal of meaning-making cultural and ideological work.[38] The "trap of appearances" that space provided, with its particular visual bias, worked by making the existing order of things, including changing relations of property and sovereignty, appear as natural, stable, or permanent.

For the first generation of U.S. colonials in the Philippines, the task of both remaking the Philippine polity and naturalizing the American colonial presence in the archipelago was a challenging one, and the results have been less enduring than Christianity's coding of sacred spaces. But the American imperial moment was nevertheless transformative—in the Philippines *and* the United States.[39] Whether in the form of prominent schoolhouses and public buildings, an "Insular" territory and government that geo-coded the archipelago at the intersection of empire and democracy, or in the cartographic knowledge production that located people, places, and things across grids of spatial intelligibility, the notion of American colonial spaces in this book is used to examine the linked ideological, material, and geographical dimensions of the production of space. The point is not that the ideological dimensions of nature, space, and landscape are of primary importance in all settings, but that questions of ideology, for Lefebvre, provided openings for linking the practical and the symbolic in the totality of space, calling attention both to the work that spaces do and the question of what they are produced for. In this sense, the notion of a *Taft-Forbes regime*, though it is not a conventional historical designation, offers both a useful periodization of American leadership in the Philippines under succeeding U.S. Republican administrations and a potentially useful scale of agency for understanding the making and coding of "American" spaces. That is, rather than supposing a unitary colonial state actor, a "regime sense of agency" in the production of space allows us to navigate between biographical and institutional lenses to highlight complex, relational, and embodied understandings of state power in a colonial context. Although by no means uncontested, the era was characterized by continuity of power among a relatively small network of actors who leveraged their positions in support of distinctive sets of priorities, even as the regime itself was necessarily dynamic in its character and constitution (notably expanding to include elite Filipino politicians), and thus in the range of ideologies that it sought to encode or realize in the Philippine landscape.

What did it mean then for the regime to produce space in this context? By tracing the development of distinctive (but intertwined) ways seeing, representing, and transforming Philippine spaces in the ensuing chapters, I attempt to work this out in the telling—in other words, we shall see. The production of space is not a ready-made framework that can be overlain without respect for context, agency, and existing relations of power, and I do not wish to push too far ahead of the arguments to be developed across the book. Some more adept storytellers may feel I have already said too much! But in emphasizing the production of space as an open (rather than teleological) intellectual framework, it is helpful to begin by stepping back from narrow economic understandings of *production* to draw instead, with Lefebvre, on Marx's reading of Hegelian production: the sense that as human beings, we are material and social beings who must produce our own existence, and in doing so, our own consciousness and our own worlds.[40] There is practically nothing, in this sense, that does not have to be produced or achieved in a fashion, as reflected in all manner of juridical, political, religious, artistic and philosophical institutions and forms. Nature itself, apprehended by the sense organs, appropriated in ideological arguments, extracted through large scale machinery, or even as something preserved, is therefore modified and in a sense produced.[41] Yet while production can be viewed broadly—as a process characterizing the reproduction of social life itself—the term still draws its rhetorical currency from a narrower process of material transformation, the production of things "in space," which is organized primarily through capitalist social relations. For as Lefebvre reminds us, there is virtually nothing in our homes, cities and their vast, planetary hinterlands that is not itself produced through relational social processes. Even the landscapes of Venice, Italy, he insists, though few would question the city's authenticity as a creative "work" rather than an industrial "product," "is a place that has been *laboured on*. Sinking pilings, building docks and harborside installations, erecting palaces—these tasks ... constituted social labor, a labor carried out under difficult conditions and under the constraint of decisions made by a caste destined to profit from it in every way. Behind Venice the work, then, there assuredly lay production."[42]

Social space, in this sense, was literally a social *product*: it required labor to be built and maintained, and its production was increasingly organized through capitalist social relations (and mediated, as Lefebvre would go on to more closely enunciate, in the emerging institutions of the modern state).[43] But while space reflected the interests of particular social and political forces, this did not mean that it could be produced at will by powerful social actors, for Lefebvre understood space as the product of contested and multi-faceted social relations. Indeed, even with the weight of the U.S. colonial establishment behind him, it would be one thing for the architect Burnham to see in Manila "the bay of Naples, the winding river of Paris, and the canals of Venice," and to envision therein, in his proposed American remodel, "a unified city equal to the greatest of the Western world, with [the] unparalleled and priceless addition of a tropical setting," and another to realize those changes on the ground in in the centuries-old Spanish colonial capital.[44] A production of space approach can keep us attuned to the priorities of the Taft-Forbes regime's vision of U.S. empire as well as to its contradictions, including core contradictions, persistently elided, of empire and democracy. Yet as Lefebvre was also at pains to show, the state itself existed in tension with social forces that persistently threatened to undermine it at "weak points," withering away an always precarious authority. These tensions were experienced acutely by late colonial states, lacking widespread legitimacy, in which the language of nationalism provided colonized peoples with a universal language of unity, development, and dissent.[45]

Colonial State Space and "American Colonial Spaces"

Before the eponymous volume, the idea of the production of space was floated in a short book, *The Survival of Capitalism*, in which Lefebvre marvels at capitalism's capacities for overcoming crises, often generated by contradictions of its own making, to continue to reproduce itself (and expand) over time. For Lefebvre, we could not "calculate at what price" the survival of capitalism was achieved, "but we do know the means: *by occupying space, by producing a space*."[46] Yet if the production of space was key to understanding capitalism's reproduction, then for some, even friendly critics like geographer David Harvey, Lefebvre had "failed to explain exactly how or why this might be the case," while others, like Don Mitchell, have posited that "Lefebvre's value is not as a theorist so much as a generative thinker adept at being suggestive."[47] And perhaps this would be useful enough. But while Lefebvre was aware of the heuristic value of the production of space as an evocative line of questioning, he also insisted on the value of rethinking space in dialectical terms as an explicitly historical concept, one that acted "retroactively upon the past, disclosing aspects and moments of it hitherto uncomprehended."[48] How did capitalism create the space of its own survival? The answer, after Lefebvre's reworking of Marxist state theory after *The Production of Space*, was to be found in the state and state spaces,[49] and in the lessons of colonization as a particular state-bound strategy of organizing territorial relations between dominant and dominated spaces.[50]

While the production of space concept allowed Lefebvre to view a range of social processes associated with the emergence of capitalism, including land commodification, urbanization, uneven development, and alienation, through a common spatial frame, it was the singular, integrative work of the state and state spaces, built around the control of particular "stocks" and "flows" and realized in a "hierarchically stratified morphology," he now argued, that made the mode of production possible.[51] At once a product and an instrument of state territoriality, violence, and empire-building, the production of state space, stimulated by war, crisis, and catastrophe, established regulations, codes, and chains of equivalence capable of reaching across geographic sites and scales. Pulling apart the relation between space and the state into three interrelated "moments"—national territory, social space, and mental space—Lefebvre placed the state at the center of a totality of power relations, a Hegelian system of systems for establishing connections among diverse elements of social life.[52] The space of national territory was itself complex, however, at once "a physical space, mapped, modified, transformed by the networks, circuits and flows that are established within it—roads, canals, railroads, commercial and financial circuits, motorways and air routes" and also "a material—natural—space in which the actions of human generations, of classes, and of political forces have left their mark, as producers of durable objects and realities."[53] While Lefebvre, in *The Production of Space*, identified the centrality of territorial conflict in European history in the making of a competitive space of capitalist accumulation, he subsequently emphasized the state's pivotal role *in* the mode of production, developed unevenly across planetary space, including the colonization and naked domination of territorial spaces by late modern states and empires.[54] As Manu Goswami has argued in the South Asian context, during the last third of the nineteenth century and first part of the twentieth, the "production of colonial state space transformed the socioeconomic geography of colonial India, consolidated modalities of state power, and deepened the reach of state-generated classificatory schemes."[55]

Viewed in this light, the extension of a reproductionist reading of the production of space from the survival of *capitalism* to the survival of *empire* is no mere side story. It gives expression to the problem of reproducing relations of power and exploitation, more broadly, in which relations of empire, race, nation, and capitalism are difficult to disentangle. Like Goswami, Stefan Kipfer and Kanishka Goonewardena have challenged the presumption that Lefebvre was unconcerned with the colonial, while at the same time arguing that Lefebvre's engagement with *colonization*, in particular, "needs to be blasted out of the Eurocentric confines of his overall work."[56] Whilst colonization first served Lefebvre largely as a metaphor evoking the dominance of capital and the state over (metropolitan) everyday life, in his turn to state theory, the language of colonization referred more explicitly to state strategies of producing space, especially the organization of territorial relations between core and periphery. As such, colonization was critical to, but also in some ways exceeded, the reproduction of capitalism:

> At the heart of 'colonisation' is also the reproduction of *relations of domination* with all the humiliating and degrading aspects that sometimes escape economistic analyses of imperialism. 'Colonisation' is *one* part of the role of the state in the reproduction of relations of production and domination: the coordination of the pulverized abstract space of capital and the thwarting of opposition with hierarchical separations of social space. Lefebvre's notion of 'colonisation' therefore alerts us to a key aspect of how the state produces abstract forms of space: homogenous, fragmented, and hierarchical.[57]

Imperialism and capitalism develop together, perfecting "methods of plunder" while offering lessons on the centrality of state interventions. Hence, "coterminous with the formation of capitalism, imperialism thus forms an integral part of original accumulation and points to a more permanent role for systemic political violence in economic processes than Marx and Engels allowed for."[58] And yet the persistence of systemic violence, in both colonial and post-colonial forms, was less a sign of strength—for capital *or* the state—than one of instability within what Lefebvre called a state mode of production.

For Goswami, colonial state space in India embodied a complex ensemble of "practices, ideologies, and state projects that underpinned the restructuring of the institutional and spatiotemporal matrices of colonial power and everyday life."[59] But consolidating economic relations around a British-centered world economy also made the subcontinent more vulnerable to famine and indebtedness, as the social differentiation and geographically uneven development intrinsic to capitalism took on exaggerated form in the Indian colonial context. Rather than a unitary space, she argues, colonial state space was a "contradictory force-field, shot through with both everyday and spectacular tensions," as practices of colonial domination engendered new forms of difference and resistance.[60] Reading the colonial state through the production of space thus offers Goswami a lens for viewing the intertwined historical geography of colonial and capitalist expansion and subsequent consolidation of the interstate system within a single analytical frame, while retaining a view that is "differentiated, multiscalar, and multitemporal."[61]

In the Philippines, of course, *American* colonial spaces were not the first such transformation. Instead, as historian Vicente Rafael has put it, the making of the modern Philippine nation-state can be traced to the successive interventions of "Spanish, North American, and Japanese colonial regimes, as well as their postcolonial heir, the Republic," wherein each, over differing temporal and spatial extents, "sought to establish power over social life, yet found themselves undermined and overcome by the new kinds of lives they had spawned. It is precisely this dialectical movement that we find starkly illuminated in the history of the Philippines."[62] The dialectical movement of empires, and imperial overreach, has also been thrown into stark relief in Philippine geographies, including the different kinds of colonial space, sometimes contested, sometimes appropriated for different ends, that I explore in this book around a far narrower time frame. Three and a half centuries after the archipelago was first cast, in an effort to charm the Habsburg heir Felipe II, as *Las Islas Filipinas*, the U.S. regime catalyzed key shifts in both colonial and postcolonial forms of state power. In focusing on the U.S. colonial project, however, it should be clear that, from a Lefebvrian perspective, the "Americanness" of colonial space in the Philippines is not taken as a *fait accompli*, even at settings like the "New Luneta" in Manila or the American "Teachers Camp" in Baguio, but rather as a partial and unstable achievement, and an ideal around which competing visions of the future were invested. And if the top-down views of empire on display at the banquet for Forbes in 1912 appeared to celebrants as perfectly natural and true, reflecting the views of a financial class coming to grips with its geographical assets, then it is clearly not enough for us to remain on Fifth Avenue, to look only through such perspectives, even though they constitute the main object of the study, without engaging with the questions that

Lefebvre compels us to consider: who produces space, for what purposes, and under what conditions?

Lefebvre argued that the "successful unmasking of things in order to reveal (social) relationships ... remains the most durable accomplishment of Marxist thought," and this reading of ideology, I have suggested, is pivotal to his development of the production of space as a historical (or historical-geographical) concept.[63] The intersection of ideology, materiality, and colonial state space in the Philippines is an important site for such an analysis, and for thinking through the evolving geographies of American empire and its limits. While the framework sketched above provides a useful set of analytical guideposts, however, it is by no means a determinist one, and does not preclude engagement with other critical ideas and frameworks as we turn with greater specificity to the production of different kinds of colonial spaces in the chapters that follow.

Design of Chapters

The book chronicles the efforts of U.S. colonial officials to extend a vision of American empire through the production of space in the Philippines in the early twentieth century. In telling that story, and in setting out the broad historical and theoretical parameters of the narrative, this introduction began, with seven courses at Sherry's on a May evening in 1912, very nearly at the end, that is, on the precipice of the demise of the political regime that had shaped the first 'long decade' of U.S. colonial policy in the Philippines. In the next chapter, we begin again by stepping back to 1898 and the onset of U.S. military intervention and colonial rule in Manila, and the book proceeds to relate an overlapping but progressive series of historical developments. Together, the chapters advance a narrative arc spanning the career of the Taft-Forbes regime in the Philippines, until the apparent sea change of the 1912 U.S. election and its aftermath.

Chapter one, "Insular territory: war, democracy, and America's 'first moment of global ambition'," narrates the origins of the U.S. occupation of the Philippines from the Spanish-American War to the Philippine-American War and concurrent establishment of the Insular Government under the U.S. War Department. Engaging with diverse understandings of the emerging American empire and colonial state formation in the archipelago, the chapter emphasizes the contested legal, political, and military construction of *Insular Territory* as a cross-cutting spatiality, worked out on the battlefield and in a series of U.S. Supreme Court cases (1901–1922), that enabled the creation of American colonial spaces over the template of more than three centuries of Spanish colonial rule. Following this broad contextualization, the ensuing three chapters then each investigate a different spatial form—map, landscape, and road—as distinctive sets of space-making practices, addressing problems of scientific, military, and governmental knowledge; architecture, aesthetics, and urban planning; and infrastructure, labor, and geographical connectivity, respectively.

Chapter two, "Map: U.S. colonial science, geo-politics, and the remapping of the Philippines," surveys the explosion of governmental mapping under the new Insular Government, as American colonial scientists, soldiers, and bureaucrats sought to turn the distant archipelago into knowable and hence governable territory. It sketches the development of a distinctive 'geo-politics' of knowledge (alongside biopolitical dimensions of colonial science and medicine) through the work of a diverse set of geographical knowledge producers, from Jesuit cartographers to Insular scientists to the paramilitary Philippine Constabulary, in their efforts to fix the position of Philippine lands, resources, and peoples within grids of power and knowledge. Drawing connections to the inheritance of Spanish geographical and ethnological knowledge, the chapter traces the emergence of a cartographic model of governmental knowledge production that worked not only to produce more accurate representations of the islands on paper but also to facilitate the extension of territorial sovereignty, and governmental techniques of calculation, at multiple sites and spatial scales.

Chapter three, "Landscape: The Burnham Plans and American landscape imperialism in Manila and Baguio" revisits a signature moment of the early colonial regime when the famed Chicago architect Daniel Burnham travelled to the Philippines (in 1904–1905) to redesign the colonial capital of Manila and, to his greater delight, produce a town plan along City Beautiful lines for the new summer capital and mountain retreat. Focusing on efforts to realize a formal vision of American colonial life in the Philippines through conventions of landscape and urban planning, the chapter traces the aesthetic dimensions of U.S. imperialism through the collaboration of Burnham and Forbes, then Commissioner for Commerce and Police, and through Forbes's continuing efforts (working with Insular architect William Parsons) to realize elements of the Burnham Plans in Manila and Baguio during his career as Governor-General. Focusing on the design of three iconic (and elite) aesthetic landscapes—the "New Luneta" on the backfilled Manila waterfront, the Plan for Baguio, and Forbes's own private "Topside" estate, located spectacularly on a ridge above Baguio—the chapter explores the implications of the regime's emphasis on distributing things "beautifully, conveniently and expediently" in space amid wider transformations of the Philippine polity, environment, and space economy.[64]

Chapter four, "Road: W. Cameron Forbes, Philippine roadwork, and the production of space," turns from the aesthetics of landscapes to the labor required to build them, and to new geographies of circulation and exchange envisioned under the regime's emphasis on "material development." While every mile of new or upgraded roadways—rendered in maps and statistics—could be celebrated by Insular officials as evidence of progress under benign American governance, each was also the product, more directly, of untold hours of backbreaking and often dangerous human toil, much of it coerced through a variety of means by the U.S. Army, Insular government, Philippine Constabulary, and local elites. The chapter explores the project of Philippine roadwork under the Taft-Forbes regime as one of both labor and geography, following Forbes's efforts in seeking solutions through the revival of the Spanish corvée, exploitation of prison labor, and constitutional strategies for ensuring a consistent labor supply. Through Forbes's signature project of roadbuilding, the chapter offers a lens onto the production of space at a granular level, while contributing to the broader picture of a coercive state apparatus intent on sweeping social and economic transformations. In Forbes's campaign for a "permanent" system of road construction, maintenance, and inspection, the chapter also explores the work that new and improved roads were expected to perform, opening pathways of markets and commerce while extending the reach of the state into heretofore more isolated places.

Chapter five, "Coda: Insular empire," serves as the book's epilogue and conclusion. It follows Forbes's travels from Manila to New York, via Moscow and Sussex, in a journey that anticipated the unraveling of the Taft-Forbes regime on the eve of the 1912 U.S. Presidential election. The election would put U.S. Philippine policy in the hands of the erstwhile anti-imperialists of the Democratic Party for the first time but did not, of course, mark the end of U.S. empire in the Philippines. Reflecting on the historically specific geographies of territory, map, landscape, and road that I attempt to reconstruct in what follows, the chapter considers the limits of the production of space as an imperial strategy but also the persistence of U.S. empire, and enduring elements of insular territoriality, in different forms.

Notes

1 W. Cameron Forbes, "Dinner in Honor of the Honorable W. Cameron Forbes ... Mounted Collection of Invitations, Menus, Typed Transcripts of Speeches, etc." (New York, 1912). W. Cameron Forbes Papers, MS 1365.1, Houghton Library, Harvard University, pp. 3–4.
2 Forbes, "Dinner." Forbes's speaking and inspection schedules during his Philippine career are closely documented in his personal journals. William Cameron Forbes, "Journal," First Series, Vol. 1–5 (1930). W. Cameron Forbes Papers, MS Am 1365, Houghton Library, Harvard University.
3 Forbes, "Dinner," pp. 4–5; Chauncey Depew in Forbes, "Dinner." Forbes noted the lack of direct steamship travel to North America (or Hawaii) under the U.S. flag and called for improvements in transportation linkages, product preparation for market, and advertising.
4 Forbes, "Dinner," p. 6. The Forbes fortune was itself built initially on the China trade from the 1830s, as the family profited on the tea and opium trades through the Russell and Company merchant firm.
5 Forbes, "Dinner," p. 6.
6 Neil Smith, *American Empire: Roosevelt's Geographer and the Prelude to Globalization* (Berkeley, CA: University of California Press, 2003); see also Smith, *The Endgame of Globalization* (New York: Routledge, 2005). I engage with Smith's notion of American empire and the significance of 1898 as the United States' "first moment of global ambition" in Chapter one.

7 "The Mayor Predicts Great Chinese War; A Dinner for Gov. Forbes" *New York Times* May 15, 1912, p. 11.
8 Ibid.
9 William Jay Gaynor in Forbes, "Dinner," p. 10.
10 Depew in Forbes, "Dinner," p. 14.
11 Forbes, "Journal," First Series, Vol. V, p. 157.
12 "Poncio Pilato" *La Vanguardia* June 25, 1912, in W. Cameron Forbes, Philippine scrapbook 1910–1913 (1930). W. Cameron Forbes Papers, fMS Am 1365.9, Houghton Library, Harvard University, p. 183 (translation mine). The author's reference to those Pontius Pilates "who know how to properly wash their hands" likely refers both to the New Testament Roman prefect of Judaea and to aggressive U.S. colonial sanitation and hygiene policies. See Warwick Anderson, *Colonial Pathologies: American Tropical Medicine, Race, and Hygiene in the Philippines* (Durham, NC: Duke University Press, 2006).
13 "Poncio Pilato" *La Vanguardia* June 25, 1912.
14 "The Forbes' Sun; El Sol de Forbes" *The Free Press* June 4, 1910, in Forbes, Philippine scrapbook. The use of Spanish dates within the scrapbook indicates that Filipino staff were employed to keep Forbes's clippings book current.
15 "Editorial: La teoría de la 'intususcepción'" *La Vanguardia* March 21, 1912, in Forbes, Philippine scrapbook.
16 Taft to Forbes, July 30, 1912. W. Cameron Forbes Papers, bMS Am 1364, File 302, Houghton Library, Harvard University; "Poncio Pilato" *La Vanguardia*.
17 Forbes to Stimson, June 17, 1912. W. Cameron Forbes Papers, MS Am 1366.1, Confidential Letter Book No. 1, Houghton Library, Harvard University.
18 Henri Lefebvre, *The Production of Space*, trans. Donald Nicholson-Smith (Oxford: Blackwell, 1991).
19 On early demands for Filipino labor under the U.S. regime, see Greg Bankoff, "Wants, Wages, and Workers: Laboring in the American Philippines, 1899–1908," *Pacific Historical Review* 74 (1998): 59–86; Justin F. Jackson, "'A Military Necessity Which Must be Pressed': The U.S. Army and Forced Road Labor in the Early American Colonial Philippines," in M.M. van der Linden and M. Rodríguez García (eds.), *On Coerced Labor* (Leiden: Brill, 2016), pp. 127–158; on Filipino-U.S. political and social dynamics during this period, see Patricio N. Abinales and Donna J. Amoroso, *State and Society in the Philippines* (Lanham, MD: Rowman & Littlefield, 2005), pp. 119–149; Filomeno V. Aguilar, Jr., *Clash of Spirits: The History of Power and Sugar Planter Hegemony on a Visayan Island* (Honolulu, HI: University of Hawaii Press, 1998), pp. 189–228; Benedict Anderson, "Cacique Democracy in the Philippines," in Anderson, *The Spectre of Comparisons: Nationalism, Southeast Asia, and the World*, pp. 192–226 (London: Verso, 1998); Michael Cullinane, *Illustrado Politics: Filipino Elite Responses to American Rule, 1898–1908* (Quezon City: Ateneo de Manila University Press, 2003); Paul A. Kramer, *The Blood of Government: Race, Empire, the United States, & the Philippines* (Chapel Hill, NC: University of North Carolina Press, 2006); Alfred W. McCoy (ed.), *An Anarchy of Families: State and Family in the Philippines* (Madison, WI: Center for Southeast Asian Studies, University of Wisconsin and Ateneo de Manila University Press, 1993).
20 Lefebvre, *Production of Space*, p. 44.
21 A useful starting point is Vicente L. Rafael, "Colonial Contractions: The Making of the Modern Philippines, 1565–1946," *Oxford Research Encyclopedia of Asian History*, June 2018. DOI: 10.1093/acrefore/9780190277727.013.268.
22 The use of the adjective "American," throughout the book, also remains problematic, but as Smith framed the dilemma, "alternatives are often awkward, and highly resonant meanings too established" to abandon the term, so I too have "retained this geographically incorrect usage." I attempt to follow my

informants in the use of the word, but refer to the government as the United States and to use the abbreviation "U.S." to indicate specific offices, institutions, and policies, where possible. Smith, *American Empire*, p. 465, n. 1.

23 Forbes to Taft, October 12, 1909, p. 2. W. Cameron Forbes Papers, MS Am 1366.1, Confidential Letter Book No. 1, Houghton Library, Harvard University.

24 Kramer, *The Blood of Government*, pp. 308–322.

25 Ibid.; on U.S. interventions in landscape during this period, see also Anderson, *Colonial Pathologies*, pp. 104–129; David Brody, *Visualizing American Empire: Orientalism & Imperialism in the Philippines* (Chicago: University of Chicago Press, 2010), pp. 140–163; Gerard Lico and Lorelei D.C. De Viana, *Regulating Colonial Spaces: A Collection of Laws, Decrees, Proclamations, Ordinances, Orders and Directives on Architecture and the Built Environment During the Colonial Eras in the Philippines* (Manila: National Commission for Culture and the Arts, 2017), pp. 107–297; Ian Morley, "Modern Urban Designing in the Philippines, 1898–1916," *Philippine Studies: Historical and Ethnographic Viewpoints* 64 (2016): 3–42; Cristina Evangelista Torres, *The Americanization of Manila, 1898–1921* (Quezon City: The University of the Philippines Press, 2010).

26 Kramer, *Blood of Government*, pp. 309–310.

27 Forbes, "Dinner," p. 7; the Governor-General's minimization of war costs was undermined, however, when Senator Depew, noted that there were $25–30 million added to the pension lists from the war in addition to $300 million in war costs. Depew, in Forbes, "Dinner."

28 Lefebvre, *Production of Space*, p. 44.

29 Henri Lefebvre, *The Survival of Capitalism*, trans. F. Bryant (New York: St. Martin's Press, 1976); Lefebvre, *Production of Space*; Lefebvre, *State, Space, World: Selected Essays*, eds. N. Brenner and S. Elden (Minneapolis, MN: University of Minnesota Press, 2009).

30 Henri Lefebvre, *Sociology of Marx*, trans. N. Gutterman (New York: Columbia University Press, 1982), p. 59.

31 Lefebvre, *Production of Space*, p. 44.

32 This lineage is discussed in Neil Smith, "Of Social Interests and Critical Intent: From Ideology to Discourse and Back Again," Progress in Human Geography Lecture, Annual Meeting of the American Association of Geographers, Seattle, WA, April 2011. Video available online: https://www.youtube.com/watch?v=H-cIVDlmdqWs, accessed June 5, 2020.

33 Lefebvre, *Sociology of Marx*, p. 60.

34 Neil Smith, *Uneven Development: Nature, Capital, and the Production of Space*, second edition (Oxford: Blackwell, 1990), p. 15.

35 Lefebvre, *Production of Space*, p. 44.

36 Ibid.

37 The notion of the concrete abstraction was central to Lefebvre's formulation of the production of space: "Production, product, labour: these three concepts, which emerge simultaneously and lay the foundation for political economy, are abstractions with a special status, *concrete* abstractions that make possible the relations of production. So far as the concept of production is concerned, it does not become fully concrete or take on a true content until replies have been given to the questions that it makes possible: 'Who produces?', 'What?', 'How?', 'Why and for whom?' Outside the context of these questions and their answers, the concept of production remains purely abstract." Lefebvre, *Production of Space*, p. 69.

38 Lefebvre, *Production of Space*, p. 73.

39 Alfred W. McCoy and Francis A. Scarano (eds.), *Colonial Crucible: Empire in the Making of the Modern American State* (Madison, WI: University of Wisconsin Press, 2009).

40 Lefebvre, *Production of Space*, pp. 68–168.
41 Lefebvre, *Production of Space*, p. 68; see also Smith, *Uneven Development*; Scott Kirsch, "Historical-Geographical Materialism" in A. Kobayashi (ed.), *International Encyclopedia of Human Geography*, Second edition, Vol. 7 (Oxford: Elsevier, 2020), pp. 31–36.
42 Lefebvre, *Production of Space*, p. 76.
43 Lefebvre, *State, Space, World*.
44 Daniel H. Burnham and Peirce Anderson, "Report on proposed improvements at Manila," Sixth Annual Report of the Philippine Commission, Part 1, pp. 627–635 (Washington, DC: U.S. Government Printing Office, 1906), p. 635.
45 Manu Goswami, *Producing India: From Colonial Economy to National Space* (Chicago: University of Chicago Press, 2004).
46 Lefebvre, *Survival of Capitalism*, p. 21, cited in Neil Brenner and Stuart Elden, "Introduction. State, Space, World: Lefebvre and the Survival of Capitalism" in Lefebvre, *State, Space, World*, pp. 1–48, p. 26.
47 David Harvey, *The New Imperialism* (Oxford: Oxford University Press, 2003), p. 87, cited in Brenner and Elden, "Introduction," p. 26; Don Mitchell, "Revolution and the Critique of Human Geography: Prospects for the Right to the City after 50 Years" *Geografiska Annaler: Series B* 100 (2018): 2–11, p. 10, fn. 3. For an earlier, related critique of Lefebvre, see Smith, *Uneven Development*, pp. 123–126. For Harvey, the problem of how capitalism secured its own survival through the production—and sometimes the destruction—of particular geographical landscapes and space economies led to a related series of analytically productive concepts, including notions of the spatio-temporal fix, creative destruction, and accumulation by dispossession.
48 Lefebvre, *Production of Space*, p. 65.
49 Brenner and Elden, "Introduction."
50 Stefan Kipfer and Kanishka Goonewardena, "Urban Marxism and the Post-Colonial Question: Henri Lefebvre and 'Colonisation'" *Historical Materialism* 21 (2013): 76–116.
51 Lefebvre, *State, Space, World*, pp. 223–253, p. 230.
52 Ibid., p. 224.
53 Ibid.
54 Lefebvre, *Production of Space*, pp. 275–282; Lefebvre, *State, Space, World*, pp. 167–306.
55 Goswami, *Producing India*, p. 9.
56 Kipfer and Goonewardena, "Urban Marxism and the Post-Colonial Question," p. 108.
57 Ibid., p. 96.
58 Ibid., p. 94. The authors attribute Lefebvre's shift from metaphors of colonization to concern for a "particular, state bound form of organizing hierarchical territorial relations" in the 1970s as a response to the anti-colonial contexts of 1968 protests, especially in connection with Vietnam and Algeria, as well as Lefebvre's engagement with Marxist theories of imperialism from Lenin and Luxemburg to contemporaries Samir Amin and Andre Gunder Frank.
59 Goswami, *Producing India*, p. 8.
60 Ibid.
61 Ibid., pp. 18–19.
62 Rafael, "Colonial Contractions," p. 1.
63 Lefebvre, *Production of Space*, p. 81.
64 Forbes, "Journal," First Series, Vol. I, p. 128.

1 Insular Territory

War, Democracy, and America's "First Moment of Global Ambition"

Founding Fictions

While the United States could not rest its claims to the Philippines, as it had for Cuba and Puerto Rico, on the basis of proximity, it was "none the less true," President William McKinley instructed U.S. negotiators in advance of the 1898 Paris peace negotiations, "that, without any original thought of complete or even partial acquisition, the presence and success of our arms at Manila imposes upon us obligations which we can not disregard."[1] It was the "march of events," McKinley explained, that "rules and overrules human action," and though it was "without any desire or design on our part, the war has brought us new duties and responsibilities which we must meet and discharge as becomes a great nation on whose growth and career from the beginning the Ruler of Nations has plainly written the high command and pledge of civilization."[2] But nor could American leaders ignore the commercial opportunities that the situation had engendered, given demands for enlarging American trade and absorbing surplus capital after the financial volatility of the 1890s. Such opportunities, as McKinley described in language echoing the views of the American naval strategist, public intellectual, and Washington insider Alfred Thayer Mahan, now depended "less on large territorial possession than upon an adequate commercial basis and upon broad and equal privileges"—namely, the "open door" to China.[3] Even as he rationalized the U.S. occupation of Manila as an accident of war and geography, however, McKinley pressed negotiators not to "accept less than the cession in full right and sovereignty of the island of Luzon."[4] Meanwhile, as the United States bargained with the Kingdom of Spain over the fate of the archipelago during the fall of 1898, the Philippine delegation, representing the Malalos Republic which controlled positions outside Manila, was not seated at the conference.

In an oft-repeated account of the events of 1898, McKinley volunteers that, before Dewey's victory at the Battle of Manila Bay, he could not have estimated the location of "those darn islands" within 2,000 miles.[5] It is a dubiously exculpatory claim to geographical "unknowing," and as with McKinley's pronouncements to a lack of foresight into historical events,

DOI: 10.4324/9780429344350-2

there is cause for skepticism. As early as June 1896, an attack on the Philippines had been included in plans for offensive warfare against Spain developed by Lieutenant William W. Kimball, a naval intelligence officer at the Naval War College in Newport, Rhode Island. While Kimball's plan focused chiefly on the "liberation" of Cuba from Spain (with some attention to "Porto Rico"), it also articulated the geographical (or geopolitical) logic of a two front, trans-oceanic war through an attack on Manila Bay, which, Kimball argued on the basis of Naval intelligence, Spain would be unable to defend.[6] Beyond an examination of short-term military prospects, Kimball's plan thus developed a rudimentary global geo-strategy for war on Spain, leaving aside tactical questions to squadron commanders, that envisioned the City and Port of Manila (or perhaps all of Luzon), but not the entire archipelago, as a site of U.S. occupation, power projection, and revenue production. As Kimball assessed in the initial white paper, which the Navy still describes as a benchmark in its evolving blueprint for war with Spain, "we could assuredly seize Manila Bay, reduce Kavite [sic.], and establish there a coaling and repairing base, and easily reduce Manila itself. With Manila in our hands it would be an easy matter to control the trade of Ilo-ilo and Cebu."[7] While Kimball's plan did not explicitly invoke the language of colonization or empire, it offered U.S. military and civilian leaders, including Assistant Secretary of the Navy Theodore Roosevelt, a clear rationale, drawing on both Mahan's and former Secretary of State William H. Seward's ideas of naval power and trans-oceanic force projection, that "Manila should be made a serious objective" of the war.[8] As Naval Intelligence tracked Spanish activity in the Pacific, and the White House followed news of the Philippine revolution with interest, plans for a two-front war gained presidential support as early as McKinley's first year in office. Whatever the extent of McKinley's geographical knowledge of the Philippines, as historian Walter LaFeber observes, he nevertheless could "judge their location closely enough to agree to Navy Department orders of December 1897, which instructed Commodore George Dewey to strike the Philippines should war occur between the US and Spain."[9]

Myths of historical happenstance and geographical ignorance, along with evolving ideologies of a reluctant American empire, would continue to find purchase in descriptions of U.S. intervention in the Philippine, but these were not the only founding fictions of the 1898 Paris peace agreement. As Article III of the treaty stipulated, for a fee of 20 million dollars Spain would cede the erstwhile *Islas Filipinas* to the United States, defining the archipelago to include all islands lying within a defined set of geographic borders. Working from British Admiralty charts published between 1856 and 1867, the U.S. Hydrographic Office "Treaty of Paris Map" is indicated at length in the treaty itself, which describes a neat cartographic line, in degrees and minutes north of the equator and east of Greenwich, extending unbroken around the archipelago and surrounding waters in a space that appears, in contrast to today's norm of ocean territories measured from

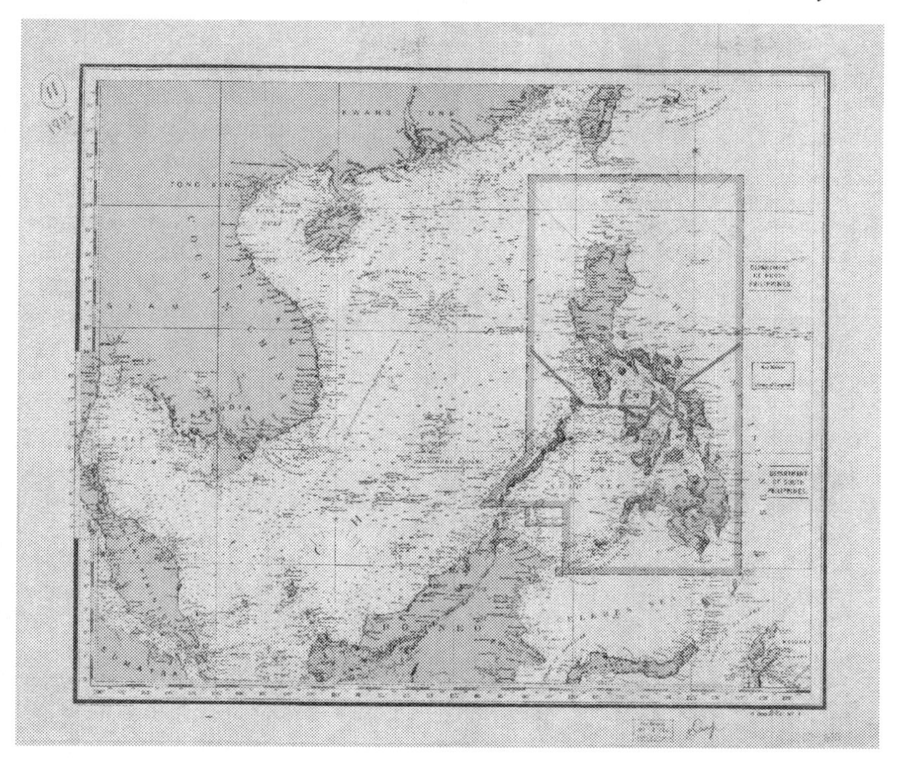

Figure 1.1 Mapping U.S. sovereignty in the Philippines within rectilinear boundaries.

Source: U.S. Government (1902). S Doc 280 57 1. Courtesy, Geography and Map Division, Library of Congress.

shorelines, in an unfamiliar form, angular and rectilinear.[10] The projection, later enhanced by a thick orange band marking the external boundary of the archipelago and a narrower blue band separating internal Departments of the North and South Philippines (see Figure 1.1), would become standard in official representations of the Philippines as a U.S. Insular Territory. Yet despite its cartographic precision, or rather, trading on the currency of that precision, it offered another founding fiction of the U.S. occupation: the idea of the map as an accurate geographical portrait of absolute sovereignty required a great deal of imaginative work. Not only, at the edges, would the extent and angles of the territory's marine borders be difficult to identify *in situ* in marine space, the map also fundamentally ignored the existence of General Emilio Aguinaldo's forces, and the first declared Philippine Republic (or Malalos Republic) over which he presided, though Filipino tactics of protracted warfare against Spain, mirroring Cuban guerrilla tactics, would present additional challenges to practical territorial sovereignty.[11] The point is not that the map of sovereignty produced in the treaty was false or unreal but that it signified the real in complex ways, both the record of a geographical transaction and an *aspirational* political cartography, marking the beginning of a wider program of territorial artifice.

In his book, *American Empire: Roosevelt's Geographer and the Prelude to Globalization*, geographer Neil Smith analyzes the shifting relations between economic and absolute (or territorial) forms of territorial expansion that characterized the rise of U.S. global power in the twentieth century. No longer, Smith argues, drawing insights from then contemporary planetary observers ranging from Halford Mackinder to Vladimir Lenin, would economic growth rest on expansion into new territories but, increasingly, in struggles over relative economic efficiency, setting the stage for an American empire that "defined its power in the first place through the more abstract geography of the world market rather than through direct political control of territory."[12] In a dazzling exposition of the "lost geographies" of the American Century, Smith describes three key "moments of global ambition," wherein the evolving American empire "grasped for global power at the beginning, middle, and end of the twentieth century," only to find its machinations, and succeeding visions of liberal empire, fall short.[13] These historical moments and their aftermaths, told through the life and work of the geographer, scientific administrator, and twice-presidential adviser Isaiah Bowman, were, nonetheless, profoundly transformative, giving rise to incipient versions of U.S. globalism and shaping the geographies of U.S. expansion as well as the shape of "extra-territorial" power. While Smith's account remains compelling, a closer engagement with the "first moment of global ambition," as it was constructed both on the map and on the insular "grounds" of the Philippines, offers a more detailed picture of changing territorial norms and practices in the wake of the Spanish-American War, with potentially different implications for understanding both the evolution of American empire and the production of colonial state space.

Smith is not wrong that recourse to colonial expansion—as a geographical solution to problems of capital overaccumulation—was not a long-term option on the scale ultimately demanded by the U.S. industrial economy. But it did come to be expressed that way in a range of, for a time, hegemonic, discourses.[14] What is more, Smith's assessment of the outcomes of 1898 for U.S. ruling class interests, that "Asian adventurism was tentative and expensive, and the colonial adventurism of 1898 garnered only the geographical crumbs of an already disintegrated Spanish Empire," may not adequately characterize what was at stake in the imperial moment, nor the special significance of the "geographical crumbs" to the production of new U.S. territorial forms.[15] In this vein, several recent reconsiderations of early twentieth century U.S. imperialism have rejected notions of the formal empire as insignificant or ineffectual because it differed from classic European models, drawing attention instead to the emergence of a complex but "agile imperial state" that was geographically fluid, and capable of exercising power in different ways in different places.[16] Whatever the islands had meant to the crumbling Spanish empire at the end of the nineteenth century, they meant something else to the United States at the turn of the twentieth century.

While relatively small in land area and population, the acquisitions of 1898, including the Philippines, Guam, Puerto Rico, effective control of Cuba, and the annexation of Hawaii, were meaningful geostrategic moves for the young republic-turned-empire, though the virtues of maintaining control of the Philippines (versus the vulnerabilities that defending the archipelago exposed) were long debated, even within the growing Naval and foreign policy establishment. The "new possessions" required (and further promoted) a heightened U.S. presence in the Caribbean and Western Pacific—militarily, politically, and commercially—and would set the stage for subsequent territorial acquisitions (and geopolitical creations) that included American Samoa in 1899 and the ten-mile-wide Panama Canal Zone in 1903, after Panama was "liberated" from Colombia with the help of American gunboat diplomacy. By the time the canal was completed in 1914, a trans-oceanic Insular Empire, comprising a network of islands and archipelagos stretching halfway around the world along tropical latitudes, was in many ways a *fait accompli* and was marked by a diversity of territorial forms, military and civilian, democratic and imperial, incorporated and unincorporated.[17] Alongside numerous interventions in Central America, lengthy military occupations of Haiti (1915–1934) and the Dominican Republic (1916–1924) would follow. Hence, for U.S. overseas empire builders, as historian Julie Greene (after Stoler) has argued, "a wide range of strategies, from formal colonization to de facto economic control, each with complex forms of graduated sovereignty, became central to its long-standing power and its ability to deny its imperial characteristics."[18] If Smith is correct to insist that, in the wake of America's first moment of global ambition, economic expansion would "henceforth bear a much more complicated relationship to geographical change," then the complexity of territorial arrangements reflected the versatility of the new U.S. imperialism, however, haphazard. The Philippines and other Insular Territories, in this sense, were neither a quaint dalliance with absolute territorial expansion nor mere missteps but rather constituted a formal (though flexible) dimension of the American liberal empire of no empire.[19]

This chapter focuses on the reproduction of the Philippines as a U.S. Insular Territory, a peculiar hybrid of military, civilian colonial, and democratic rule that would be considered both internal and external to U.S. constitutional power. In the next section, it traces the emergence of the spatial category of the *insular* (and formal category of Insular Territory) in popular, commercial, and government discourses as a particular framing of U.S. extra-territorial power at the turn of the twentieth century that reflected novel anxieties over America's place at the seat of an inter-oceanic global empire. For the relatively small cadre of authorities shaping early

U.S. Philippine policy, I suggest that the term offered an unstable resolution to the apparent contradictions of American empire and democracy. Contextualizing the American occupation during the Philippine-American War and concurrent establishment of the Insular government under the U.S. War Department, the sections that follow explore the contested legal, political, and military production of Insular Territory in the Philippines as a peculiar, cross-cutting spatiality of power—worked out on conventional and unconventional battlefields, in processes of (multiple) state formation, and in a series of U.S. Supreme Court opinions known as the Insular Cases (1901–1922)—that enabled the production of American colonial spaces over the palimpsest of centuries of Spanish rule.

Insular Affairs

> The United States never had and would never have *colonies*, only *territories*.[20]

Territory has always been a political concept; as Foucault argued, while it was "no doubt a geographical notion," it was "first of all a juridico-political one: the area controlled by a certain kind of power."[21] In this sense, as Elden concludes his account of territory as an emergent word, idea, and practice in Western political thought:

> Territory should be understood as a political technology, or perhaps better as a bundle of political technologies. Territory is not simply land, in the political economic sense of rights to use, appropriation, and possession attached to a place; nor is it a narrowly political-strategic question that is closer to a notion of terrain. Measure and control—the technical and the legal—need to be thought alongside land and terrain.[22]

If territory could be said to describe the "space within which sovereignty is exercised," it was precisely in this evolving, relational capacity that the term has carried weight in different historical and geographical settings.[23] By the time the notion of Territory itself crystallized as a westward-moving, administrative category of U.S. sovereignty in North America during the nineteenth-century, it had acquired a patina of political neutrality. For historian Richard Drinnon, the category was matched by the seemingly neutral, physical metaphor of *expansion*, rather than more analytically precise categories of colonization or empire, to describe the ongoing transformation of Western North America and the extension of U.S. empire overseas.[24] These terms, whatever their philosophical poverty, named particular forms of spatial power, and they were deployed continuously as U.S. westward "expansion" in the nineteenth century turned to increasing commercial, military, and political activity overseas before the start of the twentieth century.

The emergence of the Insular as a category of bureaucracy marked the formalization of U.S. international administration after the Spanish-American War, as the establishment of a Division of Customs and Insular Affairs (1898–1900), under the War Department, was succeeded by a Division of Insular Affairs (1900–1902) and Bureau of Insular Affairs (1902–1939). The codification of the American civil regime, from 1901, as the *Insular Government of the Philippine Islands* reflected a related rhetorical move.[25] The language of insularity provided both an accurate description of the kinds of terrestrial and marine spaces that the United States had appropriated in the Caribbean and Pacific and a physical geographic stand-in for describing, in more explicitly political terms, the relations between places that constituted the new U.S. empire.

Evolving around discourses of U.S. and British foreign policy (and naval advocacy) during the late nineteenth and early twentieth centuries, the category of the insular was at the same time acquiring a prominent place in the nascent quasi-science of geopolitics. Halford Mackinder embraced the language of *insular* and *peninsular* powers in his mapping of the "geographical pivot of history," but while, for Mackinder, the "insular crescent" would denote the oceanic zone of islands and marginal outer continents—including the Americas—surrounding a central World Island, then, for American statesmen and proto-geopoliticians like Seward, Mahan, Roosevelt, and John Hay, this was no marginal space.[26] Rather, as Denis Cosgrove describes their Imperial Pacific mappings, the "buffer zone" of the Pacific Ocean constituted precisely "the geopolitical space where East and West would finally collide" in an epic, racialized, inter-civilizational struggle.[27] And yet, while premised on a naturalized world of Darwinian competition for territory and resources, U.S. geostrategic visions for the Pacific, at least since Seward, also expressed a more nuanced view of inter-oceanic hegemony than a focus on absolute territorial distinctions might suggest, calling for a web of basing rights, coaling stations and canal passages existing in complement to formal sovereignty arrangements. For Mahan, the leading figure of the Naval War College during the 1890s (and beyond), the "natural, necessary, irrepressible," American expansion into the Pacific would provide a "nursery for commerce and shipping," and he emphasized the value of strategic bases over that of markets and resources in Pacific settings, separating the functions of empire in different geographical settings.[28] Islands and archipelagos were critical to this imagining the Pacific as an American geopolitical and commercial space, though Mahan himself, among some other Naval spokesmen and intellectuals, was ambivalent regarding U.S. occupation of the Philippines. After 1898 and the consolidation of U.S. colonialist sentiment around the Philippines, a re-situating of the archipelago's place in the Pacific—as an insular space both a part of and separate from Asia—took shape in popular and commercial mappings of the new American Pacific empire, contributing to the cartographic reconstruction of the "Philippine Islands" as a U.S. Insular Territory.

With publishers quick to capitalize on an emerging cultural imperialism in the aftermath of America's "splendid little war," and the rapid, mass production of new maps aided by advances in wax engraving technology, the popular literature was immense—in quantity and in the girth of oversized volumes that might today be called coffee table books. Large format illustrated volumes like *Our Islands and their People*, *Our New Possessions*, and *The Pearl of the Orient*, and special atlases like Rand, McNally & Co's *History of the Spanish-American War with Handy Atlas Maps and full description of Recently Acquired United States Territory*, included hundreds of pages of maps, photographs, and landscape views, while newspaper *extra* editions offered lower cost geographical illustration and reference.[29] As American mapmakers grappled with the cartographic problem of representing the Pacific as a "single geographic space,"[30] a 1902 map in *National Geographic Magazine*, featuring the Philippines at the hub of a U.S. Pacific trading empire (see Figure 1.2), embodied efforts to recast the place of archipelago in its regional and trans-oceanic relations. Positioning the Philippines as an ideal location relative to Asia—while remaining suitably offshore and insulated *from* Asia—the geopolitical and geo-economic arguments contained in the map were seen as crucial ones among U.S. colonialists.

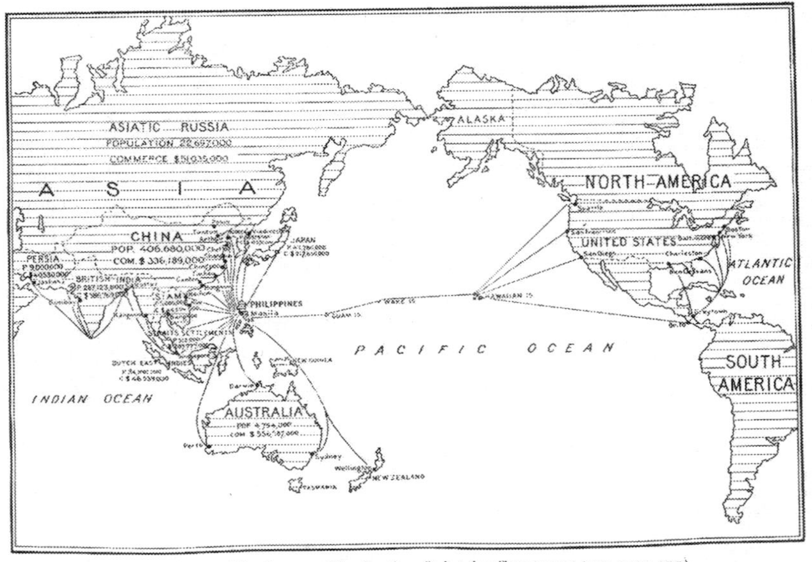

Figure 1.2 Depicting Manila as a "distributing point" for U.S. commerce in Asia.

Source: From O.P. Austin's "Problems of the Pacific—the commerce of the Great Ocean" (*National Geographic*, 1902).

The map's producer was the Chief of the U.S. Bureau of Statistics, Oscar P. Austin, by his own measure an ardent "expansionist" and then an associate editor at *National Geographic*, writing for the magazine on "Problems of the Pacific: the Commerce of the Great Ocean."[31] While numerous commentators would describe America's Pacific territorial outposts as "stepping stones" to Asia, for Austin, stressing the new, geotechnical dimensions of the imperial moment, the Hawaiian Islands, Wake Island, Guam, and the Philippines were better understood as "a continuous line of great natural telegraph poles upon which we may string a wire or series of wires, by which we may converse across this great body of water, stretching half way round the globe, making every one of its intermediate landings and relay stations on our own territory and protected by the American flag."[32] Austin includes seven full- and half-page maps of the Pacific in the 15-page article, utilizing the view from above at a small scale to promote a sense of the Pacific as a single geopolitical and geo-economic space while positioning Manila as a "distributing point for the commerce of that great semicircle of countries stretching from the Bering Strait to Australasia, containing half the population of the earth and importing a hundred million dollars' worth of merchandise every month of the year."[33] But while the *centrality* of Manila in Asia might be taken as self-evident, by *National Geographic* readers, on the basis of Austin's mappings, it did not reflect the history of the Philippines as a place decidedly at the *edge* of empires, nor the city's relatively marginal position in more recent Spanish (and British) trade networks.[34] The Manila-centric view of Asia, and fixation on the city as a "natural distributing point," were in this sense highly speculative, and the capacity to portray a world of trade relations not yet in existence was, for an audience of armchair imperialists, precisely the source of the map's rhetorical power. Later, more stylized versions of the Manila-centric projection would similarly couple the virtues of proximity to Asia with a distinctive sense of separation from it (see Figure 1.3). Meanwhile, Rand McNally's (1904) map of U.S. acquisitions, widely reproduced in contemporary textbooks and atlases, geographically juxtaposed the Insular Territories (among others) around the map of North American westward "expansion," featuring boxed inset mappings of the Philippines, Hawaii, Guam, Wake Island, Samoa, Puerto Rico, and Alaska, constructed at different scales and arrayed like satellites "in orbit" around the U.S. metropole, and naturalizing the Territories' inclusion—to an extent—in the map of the United States of America.[35]

As both a cartographic and an administrative category, the *insular* thus embodied, and helped to operationalize, a range of geographical understandings, while also reflecting novel anxieties and concerns. In a moment of escalating flows and exchanges into and out of North America, an accelerated traffic of people and things that was planetary in scale but structured significantly by national boundaries, the qualities of territory and practices of territoriality were themselves in flux. While extending influence beyond its borders, the United States depended on flows of people, goods,

Figure 1.3 *Philippine Magazine* cover (1915) positioning "Uncle Sam—Exporter" at the hub of a Manila-centric world. Land masses shaded in red ink.

Source: Courtesy, Bentley Historical Library, University of Michigan (Dean Conant Worcester Paper, 1887–1925, Box 3).

and capital for its economic growth and prosperity, making the construction of flexible or "agile" forms of legal, cultural, and political territory a matter of practical necessity. The territorial category of the Insular, embodying elements of separation and connectedness, arises in this context, but it was a complex spatiality to hold in place. As the United States extended its jurisdictions into the Caribbean and Pacific after the Spanish-American War, colonial bureaucrats, military officers, and Supreme Court justices all grappled with the question of whether the U.S. Constitution should "follow the flag" into the Insular Territories. So too did American "camp followers" and pawn shop brokers, as we will see, and Filipino nationalists and elite landowners who hoped to leverage the Constitution in support of their own democratic visions, together helping to define new forms of territory and governance at multiple spatial scales. These constructions of territory could not be imposed unilaterally but were instead worked out in complex, contested, and sometimes violent social and political relations, taking shape on unconventional battlefields, in paradoxical processes of democratization and colonial state formation, and in the courts.

Habeas Corpus

On 18 October 1901, nearly three and a half years after Dewey's Asiatic Squadron began its occupation of the Port of Manila, two seemingly contradictory impulses within the incipient American colonial state, the forces of empire and democracy, reached accommodation in an early legal skirmish that tested the efficacy of the new civil authority in the Philippines. The cases of two Americans, Harry Finnick, a pawnbroker convicted of defrauding the government in the illegal purchase of commissary stores, and Oakley Brooks, an ex-soldier working as a messenger in military service charged with refusing to obey an order (and abandoning his work contract for a more lucrative position),[36] would be of no special interest except for the conflict of jurisdictions that they brought to light after the establishment of the Insular Government. The civil government had been installed, in an American flag-draped ceremony, on the Fourth of July of the same year under the Governor-Generalship of the 43-year-old Ohio Republican, William Howard Taft. Taft had arrived in Manila the previous year as president of the Second Philippine Commission. While the First Philippine Commission (1899–1900) had been appointed, under the leadership of Cornell University President Jacob Schurman, to investigate conditions in the Philippines and to make recommendations for governance, the Second Commission (1900–1901), under Taft, was also invested with executive and legislative powers. In many parts of the archipelago, however, where the U.S. Army remained in charge (in addition to its authority in the southern archipelago), Taft's civil government was not sovereign in a practical sense. In other areas, including Luzon's upland interiors, the project of territorial sovereignty might itself be considered incomplete.[37]

At stake for the Insular Government, in the Finnick and Brooks cases, was not the territorial *extent* of U.S. sovereignty but its nature, specifically, whether military authority over prisoners could be questioned by newly appointed civil courts, i.e., *habeas corpus* protection against illegal imprisonment. Given McKinley's pledge to substitute "the mild sway of justice and right for arbitrary rule" in the Philippines, and Taft's efforts to hasten the move from war to "pacification," Finnick and Brooks could be seen as key tests of the authority of the new civil government.[38] The controversy had been triggered when the Army refused to produce Brooks for a (Philippine) Supreme Court ordered hearing, and Acting Secretary of War William Cary Sanger, in a curious abdication of authority from Washington, directed Taft and Major General Adna Chaffee, the Army's Commanding General in the Philippines, to "reach an agreement between themselves" to resolve the problem.[39] The ensuing exchange is illuminating, both for its formative character in U.S. imperialism and counter-insurgency doctrine and for the intersecting geographies of war and pacification in the Philippines that it reveals.

The problem, as Chaffee argued in a 4 October 1901 letter to Taft, was that by relinquishing *any* legal authority to the civil courts, "we arm the enemies of the United States [and] inspire them with hope."[40] He offered instead a memorandum of "propositions believed to be established as fundamental principles in a state of war, and which ought to be recognized by Civil Courts."[41] Chaffee sought to define the dimensions of the Philippine-American War in absolute spatial terms:

> A state of war exists throughout the entire Archipelago. In some parts hostilities are open and active, in other parts, inactive or the country is in a state of pacification with the Army, as we say, in observation. In all parts there may be found sentiments by some of the people, the number not known, which are opposed to the governing authority and such sentiments are always found conducive to hostile action. Therefore a state of war exists in a technical sense everywhere in the Islands.

In the context of a war that seemed to be both everywhere and nowhere, Chaffee contended, "the full power and influence of the Army, physical and moral, ought not to be lessened by the action of any Civil Court."[42]

The Army's refusal to "produce the bodies" of Finnick and Brooks may have reflected greater anxieties over the extension of liberal rights promised to *Filipinos* in the pacified provinces. As Chaffee saw it, "I do not think it is permissible in a state of war for the inhabitants in rebellion to be permitted to think that they can escape any of the consequences of a violation of the laws of war, by an appeal to the Civil Courts, which would be the case were it permitted that writs of habeas corpus could be availed in any case of arrest for such cause by military authority." He listed for Taft an array of U.S. case law references to support the contention that the "civil laws of the conquered territory" should be considered valid and binding only so far as do not impair the supremacy of the national authority. The administrative powers of the Civil Governor and Philippine Commission, Chaffee argued, must not be seen as co-extensive with those of the Commanding General under conditions of war.[43]

Taft, a trained jurist who at age 33 had become the United States' youngest solicitor general, responded to Chaffee two days later, contesting the premise that the Philippines were in a state of war throughout the archipelago—Taft limited the condition to "four or five provinces"—and challenging the absolute status that the military commander had derived from the conditions of general insurrection. Brushing aside Chaffee's case law ("The Supreme Court cases referred to in your memorandum do not seem to be in point"), Taft acknowledged that "we are still under military government in these Islands," but insisted that the power of the President, as Commander in Chief, to establish civil courts in "territory conquered or recently in a state of war" must be conceded. What was needed, Taft maintained, was "a careful working out of *concurrent jurisdiction* between the civil and the military, and the provision for the establishment of a civil government with civil liberties, all under the power of the President as Commander in Chief."[44] Taft's own status as Governor-General under the War Department (and in turn, the President), underscored the degree to which contradictions between civil and military governance were to be resolved through evolving executive powers. His notion of concurrent jurisdiction in the Philippines, wherein the expansion of liberal rights across international frontiers occurred not after military conquest but alongside it, embodied the contradictions of the imperial democracy that America was becoming.

McKinley's April 1900 Instructions to the Second Philippine Commission, calling for the inauguration of "governments essentially popular in their form as fast as territory is held and controlled by our troops," thus provided a blueprint for concurrent jurisdiction.[45] A creeping republicanism within American imperialism, limited from the outset by presumed "capacities" of the Filipino people but containing, nonetheless, horizons of autonomy and independence, mapped a dynamic, scalar framework for the emerging Philippine polity under American rule. But concurrent jurisdiction—hinging on apparently distinct spaces of war and peace—would require new sets of relations between imperial warfare and the expansion of democracy. Hence, as Taft interpreted them for Chaffee, the President's *Instructions* had called for:

> an anomalous form of military government in these Islands. In the mind of the President, the so called war which was here being conducted was peculiar and needed peculiar remedies. He evidently thought that large parts of the territory would become free from hostilities and that it would greatly aid the Army in subduing the remainder if object lessons in the benefits of American civil government could be offered to those who were still resisting the authority of the United States. He accordingly directed the Commission, first to establish municipal governments, then provincial governments and finally to recommend the form of a central government whenever the Commission should be of opinion that it could be safely transferred from Military to Civil Control.[46]

So called war? Certainly, much had changed since McKinley had pledged to Filipinos, before the outbreak of the Philippine-American War, the "full measure of individual rights and liberties which is the heritage of free peoples," and since March 1900, when Taft had been dispatched to the Philippines. While the Army had been slow to recognize the conditions of clandestine guerilla warfare into which the conflict was evolving, political conditions in Washington were in some ways unstable, even after McKinley's electoral victory in November 1900 over the anti-imperialist William Jennings Bryan foreclosed the possibility of a hasty American withdrawal from the Philippines. In September 1901, McKinley was shot at point-blank range by Leon Czolgosz while visiting the Pan-American Exposition in Buffalo, New York.[47] He died the next day, elevating Theodore Roosevelt to the presidency and leaving his own vision of democratic "object lessons" for Filipinos in the pacified provinces, as Taft had put it, largely a matter of interpretation.

It is not surprising, then, just one month into Roosevelt's term, that Washington would leave Manila to adjudicate a seemingly minor squabble over the rights of Army "camp followers" in the Philippines. Perhaps the new administration perceived that the terrain of law was one on which Taft was well suited to grapple. And perhaps it also recognized that the civil and military colonial branches were in some respects two sides of the same coin. The diffusion of democracy, in the form of establishing of municipal governments and local elections, was viewed as a key solution to the problem of Philippine "insurrection," both a tool of pacification and tactic of counter-insurgency. Taft envisioned a gradual, geographically uneven transition from war to peace, wherein the work of waging war would give way to "the suppression of insurrection and brigandage … and the maintenance of law and order."[48] This progression from war to internal policing and paramilitary operations, however, would also depend on a heterogeneous geography—characterized by simultaneous, but geographically differentiated, constellations of war and peace, even as the horizon of peace was itself premised on the extension of liberal democratic rights. As Taft argued, McKinley's promise of civil and property rights had *not* been made "with reference to governments to be established in the distant future after peace had been completely restored throughout the Archipelago; but with reference to those governments which the Commission was to establish in progress of pacification and to those laws it was to pass as the legislative branch of the Military Government." Rather, Taft insisted, "The President intended the Commission gradually to establish real civil governments in territory deemed by them to be fitted for it and gradually to relegate the army to the position which it occupies at home in time of peace, that of a power auxiliary to ordinary peace authorities and that in such real civil governments, it becomes the power and duty of the Commission to protect the individual from a violation of his civil liberty by furnishing for the purpose the usual means known to Anglo Saxon countries, that of the writ of habeas corpus in civil courts."[49]

Habeas corpus, perhaps. *Habeas territorium*, certainly. Taft's response to the Army's refusal to produce Finnick and Brooks for their hearings made clear that the extension of civil liberties—to American citizens in the first instance—was essential to the kind of imperial power that was to be expressed by the Insular Government, and to the production of territory as the "space within which sovereignty could be exercised" in the Philippines.[50] And yet it remained a patchwork model of territorial sovereignty, promising to teach and transform but not fully to internalize new subjects as it expanded. How Filipinos would respond to these object lessons in war and democracy, and attempt to leverage or re-work them for their own purposes, would in turn shape efforts to reconstruct the Philippines as both a subordinate territory and liberal democracy-in-training in the next decades. Meanwhile the question of whether, or to what extent, the U.S. Constitution "followed the flag" into the Insular Territories began percolating in a series of Supreme Court decisions that would become known as the Insular Cases (1901–1922).

Beginning with *Downes v. Bidwell*, a 1901 customs dispute in which a fruit merchant, Samuel Downes, challenged the right of customs inspectors to impose import duties on oranges brought to New York from Puerto Rico, the Insular Cases served to establish the doctrine that the Constitution did not apply *tout court* in "unincorporated" territories.[51] Rather, as Supreme Court Justice Edward Douglass White wrote in a key concurrence in the narrow decision, while the Constitution "governed the actions of the United States at any location … it was still necessary to determine the appropriate geographical scope of each constitutional provision."[52] Decisions in the Insular Cases, worked out primarily in a number of key rulings from 1901 to 1904, went a long way in delimiting the geographical scope of democracy in overseas U.S. territories, establishing, as legal scholar Efrén Rivera Ramos has argued, a set of propositions that provided the legal framework for the governance of an American colonial empire. The United States, as a sovereign nation, possessed an inherent right to acquire foreign territory, and as a corollary, that Congress possessed the right to *govern* the territory that it acquired. Majorities rested their claims on the Territory Clause of the Constitution: that "The Congress shall have Power to dispose of and make all needful Rules and Regulations respecting the Territory or other Property belonging to the United States."[53] Yet there was "a distinction to be made between something called incorporated territory and something else called unincorporated territory."[54] While incorporated territories, including the newly annexed Hawaiian Islands, were to be considered integral parts of the country, with the Constitution fully applicable, unincorporated territories, including Puerto Rico and the Philippines, were to be considered "appurtenant to, but not a part of, the United States."[55] Some but not all constitutional provisions would apply in the unincorporated territories; the Territory Clause, the Insular Cases ultimately decided, allowed Congress to determine the appropriate geographical scope of the provisions. Some Constitutional guarantees, including *habeas corpus*, were considered fundamental rights to be in extended to the unincorporated territories, while others, including rights to jury trial, and freedom from foreign customs duties, were not, bestowing on Congress the power to create around America's

Insular empire a variety of interstitial governmental boundaries.[56] Rather than creating an *extra*-constitutional zone under the category of the Insular Territories, then, the Insular decisions "were intended to provide a constitutional basis for U.S. rule over those lands."[57]

Taft responded with a gesture of compromise to Chaffee's propositions in a 17 October 1901 communication. Writing on behalf of the Philippine Commission, Taft proposed that Brooks, the former soldier, be produced in court for a scheduled hearing as to the legality of his detention, but that Finnick, the pawn broker, be remanded to prison under the existing sentence without further inquiry. The Commission would, henceforth, "forbid the issuing of a writ of *habeas corpus* against a military officer or soldier for the release of a prisoner in his custody in the provinces of Batangas, La Laguna, Tayabas, Samar, Cebu and Bohol and in all unorganized provinces."[58] The provision echoed Taft's earlier geographical claims in the extension of liberal rights in pacified versus "not-yet-pacified" territory, but Taft now pledged that the Commission would amend the code to allow for *exceptions* to the writ of *habeas corpus* in all other provinces if the Commanding General—or any military officer in command of a department or district—would certify that the prisoner was Army personnel, a prisoner of war, or a civilian employee subject to Army regulations. So, while Taft held his ground on *habeas corpus* in the pacified provinces, in straddling the apparent chasm between empire and democracy he offered Chaffee a path forward, with the Army's powers of exception built-in. Chaffee agreed to the terms the next day.[59]

Taft's doctrine of concurrent jurisdictions, further fractured into parallel state building regimes for the general and special provinces, would also provide the Insular Government with ideological cover, allowing the new civil government to cast itself not only as the alternative to military occupation but also, paradoxically, as the bearer of self-rule.[60] The extension of the "mild sway of justice and right" under Taft's governorship would take the form of an escalating, scalar expansion of electoral democracy, from the local to provincial to national scale, taking shape alongside, and in alternative to, conditions of brutal counter-insurgent warfare.

From Amigo Warfare to Cacique Democracy?

The Philippine-American war broke out in February 1899, when the U.S. Army "moved to take possession … of the territory it had procured from Spain two months earlier."[61] Following the relatively swift conventional victory over Philippine Republican forces, however, the Army struggled to occupy the territory its forces moved through easily in the first stages of the conflict.[62] High levels of exhaustion among American troops, and high rates of sickness in many units due to dysentery, malaria, and dengue fever, were among the persistent problems encountered by American forces.[63] The disappearance of visible resistance and concurrent rise of guerilla warfare constituted another. By March 1900, even as the appointment of the Second Philippine Commission was intended to usher in new forms of civil government, Army garrisons across the archipelago (see Figure 1.4) found themselves increasingly exposed to widespread, small-scale guerilla resistance.

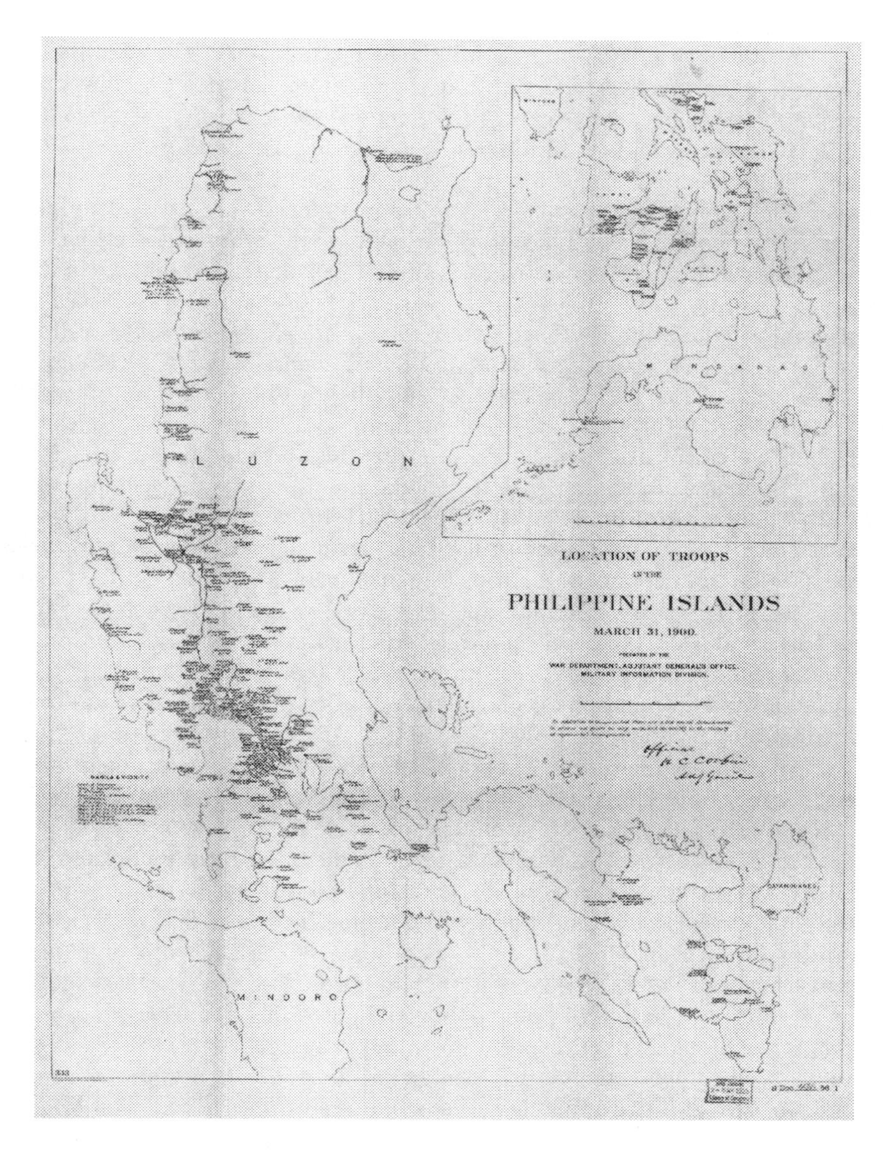

Figure 1.4 U.S. War Department (1900), "Location of Troops in the Philippine Islands, March 31, 1900." The southern archipelago is inset in the upper right of the map.

Source: Courtesy, Geography and Map Division, Library of Congress.

Insurgents initiated a range of strategies intended to drag out the war, pinning their hopes, initially, on the 1900 U.S. election in which Bryan and the Democratic Party had adopted an explicitly anti-imperialist platform. Their tactics included an emphasis on raids and ambushes, efforts to engage in combat only when guerillas possessed clear superiority, and reliance on small units that could strike and disperse, regrouping later.[64] They also relied on *sandahatan*, local village militias, to harass American communication and supply lines. Understanding the relations between guerillas and rural populations became a vexing problem for American units, as guerillas circulated between regular Philippine Republican Army and militia posts, often blending indistinguishably into villages and barrios. Where insurgents exercised power, they taxed villages for material support and sought to intimidate collaborators. So deep were the connections between insurgents and rural populations on whose support they depended that when the Army began to make significant advances in the conflict in 1901, it learned that many of the elected village mayors (or *presidentes*) were also militia *leaders*—a facet of what Americans would call *amigo warfare*.[65] Whether it was the influence of amigo warfare in the villages that constituted terrorism, as Taft would describe it,[66] or American counter-insurgent tactics of population concentration, interrogation, torture, and scorched earth campaigns, the brunt of the terror, violence, malnutrition, and disease that followed would be borne by the most vulnerable Filipinos.

While American colonials had used the term pejoratively, *amigo warfare* has been defined more generously by historian Reynaldo Ileto as "the ability to shift identities in changing contexts."[67] This mode of resistance posed particular obstacles to the larger American cavalry units searching for insurgents in the *bundok*, i.e., in remote mountainous interiors.[68] As American troops approached, guerillas could disappear into surrounding towns and countryside, donning peasant clothing. Ileto argues that many of the pacified towns were in effect under a form of dual government and that "townspeople straddled both regimes, colonial and nationalist, with relative ease."[69] Though such dual governments would be relatively short-lived, the persistence of amigo warfare in terms of shifting identities—in what Ileto describes as relations of friendship and forgetting—underscored the complexity of social and political relations in the years ahead. In the meantime, soon after McKinley's re-election placed a democratic imprimatur on America's "first moment of global ambition," assuring at least short-term continuity of U.S. Philippine policy and personnel, the Army began to implement a more strident response to amigo warfare.

In December 1900, the Army formalized an aggressive counter-insurgency strategy by instituting elements of the 1863 Lieber Code, developed to govern the Union Army's relations to local populations, including formerly enslaved people and other camp followers in the wake of advancing Union troops in the South, as well as ex-Confederate soldiers roving the countryside, during the U.S. Civil War and Reconstruction. While

Americans—from anti-imperialist critics at home to soldiers in the field—likened the Philippine-American War to the Indian Wars of the North American West, the conflict (and "post-conflict") also shared, in its practical territorial dimensions, some of the geographically piecemeal shifts in sovereignty and governance with pacification efforts in the American South during the 1860s and 1870s.[70] But while the Lieber Code had emphasized the Army's obligation to protect civilians who *accepted* American authority, in the Philippines, "General Order 100" was used to justify punishing those who did *not*, even those who had been coerced into providing material support for insurgents. While the project of pacification accelerated the restoration of municipal governments, local police, and establishment of new public schools in pacified towns, the U.S. counter-insurgency, aimed chiefly at cutting off insurgents' material bases of support, was realized in extensive destruction of crops, livestock, and villages, and in some regions, the relocation of civilians to concentration zones, camps, and villages.[71] Naval blockades were also used in efforts to starve the insurrection, reducing the capacities of peasants to share resources with insurgents. Not surprisingly, in a country that had already experienced severe crises of malnutrition during the 1896–1898 rebellion against Spain, the counter-insurgency created conditions ideal for the production of famine and diffusion of disease. While it is estimated that 22,000 Philippine soldiers would be killed in the Philippine-American War between 1899 and 1902, hundreds of thousands of civilians would perish during this time from the violence, famine, and disease that were endemic to the conflict.[72]

The brutality of the war, including Americans' use of torture and summary execution, has been well documented.[73] Pouring over news reports of official body counts, the writer and anti-imperialist Mark Twain, horrorstruck, found that only what soldiers called "the mercy of the long spoon"—the bayoneting of wounded enemy soldiers on the battlefield—could explain the high ratio of Filipino soldiers killed-to-wounded, which were nearly five-to-one over a ten-months period in 1900; he found support for the thesis in a young soldier's published letter to his mother.[74] But violence was not limited to the battlefield, and the inter-workings of war, peace, and democracy were complexly articulated in militarized projects of pacification across multiple social and geographical sites.[75] Replicating Spanish *reconcentrado* policies utilized in Cuba, the movement of civilian populations into "zones of protection" was deemed voluntary, yet those who chose not to move were to be treated as enemies or insurgent sympathizers as the Army used search and destroy missions to ravage the countryside. The Philippine Scouts, constituted by Filipinos under Army leadership, were made complicit in brutal interrogations. Meanwhile, as the counter-insurgency began to register significant gains, Army provost courts, established in areas still under martial law, "were given a free hand to try and punish suspects without evidence."[76] The provost courts also served as a tactic for disrupting the insurgents' civilian infrastructure;

through interrogations and the threat of arrest or worse, Filipinos could be leveraged to inform on guerrilla leaders.

The War Department—and its civilian Insular wing—was thus effective in establishing, through concurrent jurisdictions, a "post-conflict" colonial polity even as the counter-insurgency persisted. Municipal governments were rapidly established, particularly after the March 1901 capture of General Aguinaldo. Within four months of the erstwhile Philippine President's subsequent offer of allegiance, sovereignty over the pacified provinces had been transferred from the Army to the Philippine Commission, with Taft installed as civil governor and Chaffee as military governor retaining authority over Mindanao and the Southern archipelago (which the Army would maintain until 1913). The pace of democratization was accelerated after Taft's Fourth of July inauguration. Municipal governments were established as soon as one month after provinces were declared pacified, providing new resources for elite Filipinos in the machinery of the colonial state.[77] The offices of local democracy and the benefits of patronage politics would in this sense provide the key to a lasting peace dividend. "This peculiar consolidation of colonial rule through 'democratic means,'" as Abinales and Amoroso have described it, "hastened the conversion of Filipino elites to the American side. Once they had judged the war a lost cause, they looked for a way to come out of it with their wealth and status intact. One clear way was to take up the American offer to help govern."[78]

Alongside these extensions of democracy and patronage, insurgency and counter-insurgency persisted, particularly in the Eastern Visayas and Southern Luzon.[79] On Samar, where support for *insurrectos* had been maintained, for a time, through acquisition of food and capital from British trading houses intent on keeping the island's hemp trade open,[80] a pre-dawn attack killed 48 American soldiers from an infantry regiment of 76, wounding most of the others, and the "Balangiga Massacre," as it became known among Americans, provided military leaders with the impetus and the rationale for expanding the war. Just a week before Chaffee had warned Taft against the extension of *habeas corpus* rights throughout the Philippines, the Army had responded to the surprise attack on its troops with a campaign of unparalleled brutality in Samar and southern Luzon in the Fall and Winter of 1901–1902, razing villages, barrios, and crops to the ground in a scorched earth campaign that would make Samar—and the Philippine-American War more broadly—virtually synonymous with war atrocity in American public memory, particularly after controversial U.S. Senate and court martial hearings in 1902.[81] The war in Samar, where perhaps 10% of the island's population had been reconcentrated in camps or cordoned in zones, also made conditions of overcrowding, poor sanitation, and food shortages more acute, setting the stage for the devastating cholera epidemic.[82] Small wonder that Taft found it prudent to clarify the distinctiveness of the civil public sphere under his administration at this historical moment—the onset of the Samar campaign—whatever the Army's powers of exception.

On 4 July 1902, President Roosevelt proclaimed from Washington that the "Philippine Insurrection" was over, declaring a general amnesty and pardon for insurgents willing to pledge their loyalty to the United States. If Roosevelt's claim that "peace has been established in all parts of the archipelago except in the country inhabited by the Moro tribes" was premature, another founding fiction of American colonial spaces in the Philippines, it was not exactly wrong; the truce, and self-conscious shift from the prosecution of war to the policing of banditry, brigandage, and caciquism, was aspirational, a matter of cultural politics as well as military prowess.[83] Roosevelt's Fourth of July peace initiative reflected the consolidation of the support of Philippine military leaders, wealthy *ilustrados*, and other provincial *principales*, achieved largely through generous terms of surrender for elite classes. By 1903, with the assistance of elite Filipino families able to leverage the expansion of local municipal governments and democratic elections as a means of entrenching their own power, the United States had (re)established 1,035 municipal governments and 31 provincial governments in the Philippines under the authority of the Insular government.

Benedict Anderson has argued that, during this time, local Filipino political bosses (so-called *caciques*) took advantage of the proliferation of local and provincial elective offices, in the *absence* of autocratic territorial control or a geographically widespread bureaucracy, to entrench themselves in power in a decentralized oligarchy that would persist into the Commonwealth and post-colonial periods. This development of what Anderson calls *cacique democracy* would be abetted by American colonial land policies—including those organized around the breakup of the Spanish Friar estates—which resulted in greater agglomeration of land under wealthy Filipinos.[84] Anderson's use of the word *cacique*, however, which McCoy defines as "a term American colonials borrowed from Latin America and applied, with clear pejorative intent, to the Filipino local officials who combined local office with land and economic resources," has been criticized for replicating American colonial perspectives and pejoratives.[85] But in this sense, keeping in mind the analytically sharp, relational, and geographical sense of power that the notion of cacique democracy under U.S. rule helps to illuminate, the term also remains useful in shedding light on the American colonials themselves in their efforts to understand the Philippine polity, and the growing sense, which would become pervasive in the Taft-Forbes regime, of an empire "spoiled" by democracy, or by Filipinos themselves.

Alongside the tangled processes of war, pacification, and democratization, the American colonial state in the Philippines emerged with a complex, overlapping administrative structure in what were called, after Spanish precedent, the regular and special provinces. In pacified provinces, the Insular Government held authority over an estimated six million Filipinos categorized as "Christianized" or "civilized," along with the indigenous peoples, numbering 600,000-to-700,000 people, living in the Cordillera of northern Luzon. The Special Provinces would include both the Cordillera

Province (renamed *Mountain* Province in 1908), in the mountainous interior of northern Luzon under the Insular Interior Department (with Philippine Constabulary forces), and the Moro Province, which included about 800,000 "Mohammedans" under Army rule. The Special Provinces, constituting roughly 40% of land area in the archipelago, were understood as rich in resources and relatively sparsely settled, would be seen by American colonials both as ideal sites for the marriage of Philippine resources with American technology and capital and as key settings for experimentation with new forms of colonial governance. Insular officials would jealously guard them from what they saw as encroachments from the civil regime, especially after establishment of the Philippine Assembly, acting as an elected "lower house" to the American-dominated Philippine Commission, in 1907.

While the United States borrowed key designations—and institutional geographies—from its colonial predecessor, in other ways it departed from Spanish norms, particularly around expanding Filipino participation in democratic governance and civil service and in building a universal system of public education in the metropolitan language.[86] Performing the role of liberators against Spanish despotism had provided ideological currency for the Spanish-American War, and the ruling cliques in Washington and Manila remained committed to an exceptionalist American imperialism that required that the colonial state be seen as a progressive one, characterized by political and cultural tutelage, educational reform, and what was seen as enlightened governance. Democratization was initially a class-based colonial intervention, however. As written into the 1901 civil code for municipal and provincial elections enacted by the Philippine Commission, the right to vote was reserved for Filipinos who were male, at least 23 years of age, and could prove to have resided in a given municipality for at least six months. But the electors were also required to belong to one of three additional categories: individuals who could speak, read, and write Spanish or English; individuals who owned real property worth a specified value; and individuals who held local government positions prior to 1899. Given the paucity of public education offered under Spain, the first qualification was, initially, scarcely less restricted by class than the second, but this would change as the Americans expanded English language public education.

Like the diffusion of democracy into recently pacified towns, the U.S.-lead expansion of public education began on the precipice of the most violent phase of the Philippine-American war. In August 1901, the first boatload of American teachers had arrived in Manila Bay to be dispersed throughout the pacified provinces, in some cases replacing soldiers who had constituted the first wave in the mass literacy campaigns that were considered critical to the creation of a democratic polity in the Philippines.[87] Additional opportunities for Filipino advancement were to be found in the civil service. By 1903, Filipinos occupied nearly half of all civil administration positions, later surpassing 90% after the Taft-Forbes Regime network of senior

American bureaucrats, for whom supposed Filipino incompetence (and racial inadequacy) remained a powerful colonial *raison d'être*, faded from power in the Philippines after the American Presidential election of 1912.[88] Meanwhile, the achievements of counterinsurgency could be measured in the drawing down of the American troop presence. From a peak of 108,800 troops, the U. S. reduced its forces to 72,000 soldiers by the middle of the decade and to 13,000 by 1910, as the role of the Philippine Constabulary was elevated in policing and military (or paramilitary) activities.[89] The problem of popular insurgency in the Philippines, reproduced through conditions of landlessness, poverty, and abusive social relations, was by no means fully resolved during this time, but conditions of low-level or banal imperial warfare did reflect an essentially stable political dynamic for the "dual" colonial state until 1913, when the division between military and civilian governments was collapsed into a single process of state formation under the Insular Government in Manila.

Toward an Empire of No Empire

> There must be two Americas: one that sets the captive free, and one that takes a once-captive's new freedom away from him, and picks a quarrel with him with nothing to found it on; then kills him to get his land.[90]

Responding to what seemed the sudden emergence of a U.S. inter-oceanic empire after 1898, the founders of the American Anti-Imperialist League made the case, in a widely circulated platform statement, that imperialism was fundamentally incompatible with democracy. Since "governments derive their just powers from the consent of the governed," the group insisted, "subjugation of any people is 'criminal aggression' and open disloyalty to the distinctive principles of our government." The Anti-Imperialists reflected an ideologically diverse single-issue movement—its members included not only prominent moral critics like Twain but embodiments of both capital and labor in the personages of Andrew Carnegie and Samuel Gompers—that had made common cause in the abhorrence of classical, European-style imperialism, rejecting outright the "extension of American sovereignty by Spanish methods... The United States have always protested against the doctrine of international law which permits the subjugation of the weak by the strong. A self-governing state cannot accept sovereignty over an unwilling people."[91] By the time of Twain's chilling observation of *two* Americas, however, published in the *North American Review* in 1901, the accommodation of empire and democracy, though "curious" to his "Person Sitting in Darkness" and odious to Twain himself, seemed entirely plausible.

Narratives of the construction and manifestation of Insular Territory in the Philippines, presented in this chapter, may offer something of a geographical corrective to the well-intentioned exceptionalism of the Anti-Imperialist League, which rested on the absolute opposition of empire and democracy. For rather than acting as a restraint on imperialism, democracy provided a tool for American empire builders, at once an alternative to war, for (re)subjugated Filipinos, and a means of entrenching collaborative interests around particular relations and distributions of power. Democracy was *internal* to American empire in the Philippines, not separate from it: articulated in a space of "concurrent jurisdictions" and shifting gradations of institutional, democratic, and class-based authority, not an absolute space coterminous with a map of U.S. sovereignty. If, as I argued in the introduction, empire persevered by producing the space(s) of its own survival, then American innovations in the production of territory—and Insular Territory—as a space in which different kinds of sovereignty could be exercised, constituted pivotal moments in the making of U.S. overseas power after the Spanish-American War, lying at the crux of Twain's two Americas, liberator, and conqueror. But the American "empire of no empire" remained a product of violence, and its colonial spaces in the Philippines could persist only as unstable achievements, reinforcing existing power relations while also generating new contradictions and problems of governance in a world wherein conditions of banal insurgency and counterinsurgency were to be tolerated alongside—and framed in opposition to—the extension of democratic rights, education, and individual liberties. How the Insular Government endeavored to *know* the islands it presumed to govern is the subject of the next chapter.

Notes

1 William McKinley, "Instructions to the Peace Commissioners," September 16, 1898, Washington. Office of the Historian (U.S. State Department), Papers Relating to the Foreign Relations of the United States, with the Annual Message of the President Transmitted to Congress, December 5, 1898, Document 776, https://history.state.gov/historicaldocuments/frus1898/d776.
2 Ibid.
3 Ibid.
4 Ibid. Terms for Spain's relinquishment of claims over Cuba, Puerto Rico, and an unnamed island (Guam) in the Ladrones (Mariana Islands) had largely been agreed to in a prior armistice agreement.
5 See, for example, Walter LaFeber, *The New Empire: An Interpretation of American Expansion 1860–1898* (Ithaca, NY: Cornell University Press, 1998), p. 361; Paul S. Boyer, et al., *The Enduring Vision: A History of the American People* (Boston, MA: Cengage Learning, 2012), p. 493. Neil Smith describes the story as perhaps apocryphal in *American Empire: Roosevelt's Geographer and the Prelude to Globalization* (Berkeley, CA: University of California Press, 2003), p. 1.
6 Lieutenant William W. Kimball, "Plan of Operations Against Spain" June 1, 1896. Extract available: https://www.history.navy.mil/content/history/nhhc/research/publications/documentary-histories/united-states-navy-s/pre-war-planning/plan-of-operations-a.html.
7 Ibid., p. 13.
8 Ibid., p. 14.
9 LaFeber, *New Empire*, p. 361.

10 The basis of the Treaty of Paris map on British Admiralty charts is noted in *The Murillo Bulletin* 1 (2016), p. 3. On the cartographic construction of the Philippines and its afterlives in contemporary boundary disputes with China, see Dylan Michael Beatty, "Re-inscribing Propositions: Historic Cartography and Philippine Claims to the Spratly Islands," *Territory, Politics, Governance* 9 (2020): 434–454; from an international law perspective, see also Lowell B. Bautista, "The Historical Context and Legal Basis of the Philippine Treaty Limits," *Asian-Pacific Law & Policy Journal* 10 (2008): 1–31.

11 See Brian McAllister Linn, *The U.S. Army and Counterinsurgency in the Philippine War, 1899–1902* (Chapel Hill, NC: University of North Carolina Press, 1989).

12 Smith, *American Empire*, p. 19.

13 Smith, *American Empire*, p. xvii. See also Neil Smith, *The Endgame of Globalization* (London: Routledge, 2005).

14 LaFeber traces the development of these discourses in terms of their intellectual, strategic, and economic formulations in *New Empire*, pp. 62–196.

15 Smith, *American Empire*, p. 16.

16 McCoy et al. trace the formation of an "agile, transnational imperial state," explicitly reframing the "view of the United States, in its overseas operations, as an agile state whose diffuse, delegated power has been the source of a surprising resilience," in Alfred W. McCoy, Francisco A. Scarano, and Courtney Johnson, "On the Tropic of Cancer: Transitions in the U.S. Imperial State" in McCoy and Scarano (eds.), *Colonial Crucible: Empire in the Making of the Modern American State* (Madison, WI: University of Wisconsin Press, 2009), 3–33, p. 3, 24. Julie Greene describes U.S. empire as "changeable, fluid, and profoundly movable," and exerting power "in different ways in different places" in Greene, "Movable Empire: Labor, Migration, and U.S. Global Power during the Gilded Age and Progressive Era," *Journal of the Gilded Age and Progressive Era* 15 (2016): 4–20, p. 5. Others have emphasized continuities, amid these imperial dynamics, between settler colonialism in North America and "extra-territorial" U.S. imperialism from 1898; see Richard Drinnon, *Facing West: The Metaphysics of Indian-Hating and Empire Building*, 2nd ed. (Norman, OK: University of Oklahoma Press, 1997), Alyosha Goldstein (ed.), *Formations of United States Colonialism* (Durham, NC: Duke University Press, 1914); McCoy and Scarano (eds.), *Colonial Crucible*; Lanny Thompson, "The Imperial Republic: A Comparison of the Insular Territories under U.S. Dominion after 1898," *Pacific Historical Review* 71 (2002): 535–574.

17 McCoy et al., "On the Tropic of Cancer."

18 Greene, "Movable Empire," p. 6; Laura Ann Stoler, "On Degrees of Imperial Sovereignty," *Public Culture* 18: 1 (2006), pp. 125–146.

19 Smith, *American Empire*, p. 14.

20 Drinnon, *Facing West*, p. 295.

21 Michel Foucault, "Questions on Geography" (interview with *Hérodote* January 1976), in C. Gordon (ed.), *Power/Knowledge* (New York, NY: Pantheon, 1980), 63–77, p. 68.

22 "What is crucial in this designation," Elden insists, "is the attempt to keep the question of territory open. Understanding territory as a political technology is not to define territory once and for all; rather it is to indicate the issues at stake in grasping how it was understood in different historical and geographical contexts." Stuart Elden, *The Birth of Territory* (Chicago, IL: University of Chicago Press, 2013), pp. 322–323.

23 Ibid., p. 329. See also Stuart Elden, "Governmentality, Calculation, Territory," *Environment and Planning D: Society and Space* 25 (2007): 562–580; Christian C.

Lentz, *Contested Territory: Điện Biện Phủ and the Making of Northwest Vietnam* (New Haven, CT: Yale University Press, 2019).

24 Drinnon, *Facing West.*

25 The Bureau's post-Second World War successor agency, the Office of Insular Affairs, continues to coordinate U.S. federal policy in the Territories of American Samoa, Guam, U.S. Virgin Islands, and the Commonwealth of the Northern Mariana Islands, and to administer U.S. federal programs in the Republics of the Marshall Islands, Micronesia, and Palau. It operates under the U.S. Department of the Interior.

26 Halford J. Mackinder "The Geographical Pivot of History," *The Geographical Journal* 23 (1904): 421–437.

27 Denis E. Cosgrove, *Geography & Vision: Seeing, Imagining and Representing the World* (London: I.B. Tauris, 2008), 185–202, p. 190; See also Gearoid O'Tuathail, *Critical Geopolitics* (Minneapolis, MN: University of Minnesota Press, 1996), pp. 38–43; James Tyner, *America's Strategy in Southeast Asia* (Boulder, CO: Rowman & Littlefield, 2007), pp. 1–26.

28 In LaFeber, *New Empire*, pp. 92, 90–91.

29 W.S. Bryan (ed.), *Our Islands and their People as seen with Camera and Pencil* (St. Louis, MO: N.D. Thompson Publishing Co, 1899); G.W. Browne, *The Pearl of the Orient* (Boston, MA: Dana Estes and Co, 1900); T. White, *Our New Possessions* (Chicago, IL: Monarch Book Co, 1898); Rand, McNally & Co, *History of the Spanish-American War with handy atlas maps and full description of Recently Acquired United States Territory*, Geographical Series Vol. 11 (29 September 1898), (Chicago, IL: Rand, McNally & Co, 1898); New York Tribune, *Our New Possessions and the diplomatic processes by which they were obtained: with maps and portraits*, Library of Tribune Extras Vol. XI (June 1899) No. 6 (New York, NY: The Tribune, 1899).

30 Cosgrove, *Geography & Vision*, p. 185.

31 O.P. Austin, "Problems of the Pacific: The Commerce of the Great Ocean," *National Geographic Magazine* 13 (1902): 303–318; he identifies himself as an expansionist in O.P. Austin, *Steps in the Expansion of our Territory* (New York, NY: Appleton & Co, 1903).

32 Austin, "Problems of the Pacific," pp. 313–314.

33 Ibid., p. 316.

34 "Thanks to an accident of geography," as Vicente Rafael describes it, "the archipelago was at the farthest margins of the spread of Hinduism, Buddhism, and Confucianism," with Islam arriving as late as the twelfth century, and Spanish colonial rule, given the distance from Spain and the absence of large reserves of precious metals, serving to position the Philippines more as imperial outpost than settler colony. Vicente L. Rafael, "Colonial Contractions: The Making of the Modern Philippines, 1565–1946," *Oxford Research Encyclopedia of Asian History*, June 2018. DOI: 10.1093/acrefore/9780190277727.013.268, p. 2.

35 Raymond D. Craib and D. Graham Burnett, "Insular Visions: Cartographic Imagery and the Spanish-American War," *The Historian* 61 (1998): 101–118; Susan Schulten, *A History of America in 100 Maps* (Chicago, IL: University of Chicago Press, 2018), pp. 170–171.

36 *Army and Navy Journal* October 12, 1901, p. 125.

37 Rebecca Tinio McKenna describes the mountainous interior territories of northern Luzon acquired by the U.S. in 1898 as not entirely "stateless," along the lines of James C. Scott's anarchist history of upland Southeast Asian peoples, but existing in limited exchange with the state in a world wherein wealth, trade, and geographical isolation offered potential means of maintaining relative independence for upland peoples. Rebecca Tinio McKenna, *American Imperial*

Pastoral: The Architecture of US Colonialism in the Philippines (Chicago, IL: University of Chicago Press, 2017); James C. Scott, *The Art of Not Being Governed: An Anarchist History of Upland Southeast Asia* (New Haven, CT: Yale University Press, 2009).

38 McKinley's Benevolent Assimilation Proclamation of December 21, 1898 was published in the Philippines on 4 January 4, 1899.
39 *Army and Navy Journal* October 12, 1901, p. 125.
40 Chaffee to Taft, October 4, 1901. Worcester Philippine Collection, Vol. 1, file 30. Special Collections Library, University of Michigan.
41 A.R. Chaffee, "Memorandum" October 11, 1901. Headquarters, Division of the Philippines (El 5476-al). Worcester Philippine Collection, Vol. 1, File 30. Special Collections Library, University of Michigan.
42 Ibid.
43 Ibid.
44 Taft to Chaffee, October 13, 1901. Ex. B. File No. 5476. Worcester Philippine Collection, Vol. 1, File 30. Special Collections Library. University of Michigan. Emphasis added.
45 In Patricio N. Abinales and Donna J. Amoroso, *State and Society in the Philippines* (Lanham, MD: Rowman & Littlefield Publishers, 2005), p. 136.
46 Taft to Chaffee, October 13, 1901.
47 McKinley's assassination, as Benedict Anderson points out, was no singular event but part of a string of at least 13 major political assassinations worldwide between 1894 and 1913, mainly by anarchists like Czolgosz, that were fundamentally transforming the nature of war and peace and their spatialities. Anderson, *The Spectre of Comparisons: Nationalism, Southeast Asia and the World* (London: Verso, 1998), p. 200.
48 Taft to Chaffee, October 13, 1901.
49 Ibid.
50 Elden, *Birth of Territory*, p. 329.
51 Downes v. Bidwell 182 U.S. 244 (1901). The doctrine remains in force.
52 In Gerald L. Neuman, "Introduction," in Neuman and Brown-Nagin (eds.), *Reconsidering the Insular Cases: The Past and Future of the American Empire* (Cambridge, MA: Human Rights Program at Harvard Law School, 2015), p. xiii.
53 United States Constitution, Article IV, Section 3, Clause 2.
54 Efrén Rivera Ramos, "The Insular Cases: What is There to Reconsider?" in Neuman and Brown-Nagin, *Reconsidering the Insular Cases*, pp. 29–59, p. 31.
55 Ibid.
56 It is notable that not until the Payne-Aldrich Act of 1909 were conventional colonial 'free-trade' relations largely established between the United States and Philippines.
57 As Rivera Ramos continues, the Insular decisions thus "legitimized, via constitutional argument, the possibility of an indefinite condition of political subordination. In that sense, the Insular Cases put the US Constitution at the service of colonialism." Rivera Ramos, "The Insular Cases," p. 35, after Christina Duffy Burnett, "Untied States: American Expansion and Territorial Deannexation," *University of Chicago Law Review* 72 (2005): 797–879. Or, as Joaquín Villanueva has described it in the Puerto Rican context, "The Court ruled that Puerto Rico belonged to, but was not part of the United States." Villanueva, "The Criollo Bloc: Corruption Narratives and the Reproduction of Colonial Elites in Puerto Rico, 1860–1917," *CENTRO Journal* (forthcoming).
58 Taft to Chaffee, October 17, 1901. Worcester Philippine Collection, Vol. 1, file 30. Special Collections Library, University of Michigan.

59 Chaffee to Taft, October 18, 1901. El 5476-al. Worcester Philippine Collection Vol. I, file 30. Special Collections Library, University of Michigan. The last of the Insular Cases, *Balzac v. Porto Rico* (1922), a unanimous opinion reaffirming the constitutionality of the ad hoc nature of constitutional rights in unincorporated territories, would be presided over by then Chief Justice Taft.
60 On dual state building in the Philippines and its impacts, see Patricio N. Abinales, "Progressive-Machine Conflict in Early-Twentieth-Century U.S. Politics and Colonial State Building in the Philippines," in J. Go and A.L. Foster (eds.), *The American Colonial State in the Philippines: Global Perspectives* (Durham, NC: Duke University Press, 2003), pp. 148–181.
61 Reynaldo C. Ileto, "The Philippine-American War: Friendship and Forgetting," in A. Velasco Shaw and L.H. Francia (eds.), *Vestiges of War: The Philippine-American War and the Aftermath of an Imperial Dream, 1899–1999* (New York: New York University Press, 2002), pp. 3–21, p. 4.
62 Brian McAllister Linn, *The U.S. Army and Counterinsurgency in the Philippine War, 1899–1902* (Chapel Hill, NC: University of North Carolina Press, 1989).
63 Ibid.; Warwick Anderson, *Colonial Pathologies: American Tropical Medicine, Race, and Hygiene in the Philippines* (Durham, NC: Duke University Press, 2006), pp. 13–44.
64 Linn, *U.S. Army and Counterinsurgency.*
65 Ileto, "The Philippine-American War."
66 Linn, *U.S. Army and Counterinsurgency*, p. 18.
67 Ileto, "The Philippine-American War," p. 7.
68 Tagalog for *mountain*, the term *bundok* also connoted a broader sense of the remote mountain spaces from which the American English *boondocks*, as remote or isolated country, is derived.
69 Ileto, "The Philippine-American War," p. 7.
70 Many of the Army regulars and officers had been trained in skirmish techniques from the late nineteenth century 'Indian Wars' in the North American West, while some senior officers were also veterans of the Civil War or Union occupation of the South (i.e., Reconstruction). Drinnon, *Facing West*; Linn, *U.S. Army and Counterinsurgency*; David J. Silbey, *A War of Frontier and Empire: The Philippine-American War, 1899–1902* (New York, NY: Hill and Wang, 2007). See also Eric Foner, *Reconstruction: America's Unfinished Revolution, 1863–1877*, Parkman Prize edn. (New York, NY: Harper & Row, 2005); Scott Kirsch and Colin Flint, "Geographies of Reconstruction: Rethinking Post-War Spaces" in M. Turner and F.P. Kühn (eds.), *The Politics of International Intervention: The Tyranny of Peace* (London: Routledge, 2015), pp. 39–58.
71 William N. Holden, "The Samar Counterinsurgency Campaign of 1899–1902: Lessons Worth Learning?" *Asian Culture and History* 6 (2014): 15–30; Linn, *U.S. Army and Counterinsurgency*; Reynaldo C. Ileto, *Knowledge and Pacification: On the U.S. Conquest and the Writing of Philippine History* (Quezon City, Philippines: Ateneo de Manila University Press, 2017), pp. 3–126.
72 Abinales and Amoroso, *State and Society in the Philippines.* Civilian casualty statistics remain contested. Abinales and Amoroso, based on the findings of the Philippine–American War Centennial Initiative, estimate broadly that more than 500,000 civilians died under such conditions between 1899 and 1902 in Luzon and the Visayan Islands, including the effects of the cholera epidemic of 1902, and about 100,000 more dead in Mindanao and the southern archipelago. Official U.S. casualty estimates, in addition to "over 20,000 Filipino combatants," ascribe the deaths of "as many as" 200,000 civilians to violence, famine, and disease during the war. U.S. State Department, Office of the Historian, "The Philippine-American War, 1899–1902" (no date, accessed December 5, 2021): https://history.state.gov/milestones/1899-1913/war, while

J.M. Gates, "War-related Deaths in the Philippines, 1898–1902," *Pacific Historical Review* 53 (1984): 367–378, attributes 34,000 deaths (among the "non-Muslim Filipino population") to direct results of the war, and up to 200,000 dead from cholera. Gates notes that about 4,200 American soldiers died in the Philippines out of some 125,000 who served there during the war.

73 See R.E. Welch, Jr. "American Atrocities in the Philippines: The Indictment and the Response," *Pacific Historical Review* 43 (1974): 233–253; D.R. Smith, "American Atrocities in the Philippines: Some New Evidence," *Pacific Historical Review* 55 (1986): 1–3; Paul Kramer, "The Water Cure: Debating Torture and Counterinsurgency a Century Ago," *The New Yorker* 25 February 2008: 38–43.

74 Based on the reports of General Arthur MacArthur, Twain specified that Filipino losses during this period of 1900 were 3,447 killed, 694 wounded; during this time Americans suffered 268 soldiers killed, 750 wounded. Mark Twain, "To the Person Sitting in Darkness," *North American Review* February 1901: 161–176.

75 As Ileto observes, Americans in the pacified provinces, occupying the protected zones around garrisons, church, and town centers, performed a striking *recolonization* of the traditional Spanish colonial landscape. "The Philippine-American War."

76 Linn, *U.S. Army and Counterinsurgency*, p. 25.

77 For Americans, the best way to ensure stability was thus "to surrender large portions of the state to powerful Filipinos who had formerly resisted the U.S. invasion." Paul A. Kramer, *The Blood of Government: Race, Empire, the United States, & the Philippines* (Chapel Hill, NC: University of North Carolina Press, 2006), p. 171.

78 Abinales and Amoroso, *State and Society in the Philippines*, p. 125. See also Alfred W. McCoy (ed.), *An Anarchy of Families: State and Family in the Philippines* (Madison, WI: Center for Southeast Asian Studies, University of Wisconsin and Ateneo de Manila University Press, 1993); Michael Cullinane, *Illustrado Politics: Filipino Elite Responses to American Rule, 1898–1908* (Quezon City: Ateneo de Manila University Press, 2003).

79 Linn, *U.S. Army and Counterinsurgency*; on the project of pacification in the Muslim southern archipelago, and its evolution as a civilizational project under direct U.S. military rule without equivalent trappings of democracy, see Oliver Charbonneau, *Civilizational Imperative: Americans, Moros, and the Colonial World* (Ithaca, NY: Cornell University Press, 2020).

80 Holden, "Samar Counterinsurgency Campaign."

81 Welch, Jr. "American Atrocities in the Philippines."

82 The estimate is from A.W. McCoy, *Policing America's Empire: The United States, the Philippines, and the Rise of the Surveillance State* (Madison, WI: The University of Wisconsin Press, 2009) cited in Holden, "Samar Counterinsurgency Campaign."

83 Theodore Roosevelt, "Proclamation 483—Granting Pardon and Amnesty to Participants in Insurrection in the Philippines," July 4, 1902, W (available online: https://www.presidency.ucsb.edu/documents/proclamation-483-granting-pardon-and-amnesty-participants-insurrection-the-philippines).

84 Benedict Anderson, "Cacique Democracy in the Philippines," in Anderson, *The Spectre of Comparisons: Nationalism, Southeast Asia and the World* (London: Verso, 1998), pp. 192–226.

85 McCoy, *Anarchy of Families*, 537; Reynaldo C. Ileto, "The Centennial of 'Cacique Democracy': Constructing Politics in a Time of Pacification" (2014) in *The Adrian Cristobal Lecture Series 2010–2017* (Manila, 2018), pp. 91–114.

86 Abinales and Amoroso, *State and Society in the Philippines*, pp. 119–127; Anderson, "Cacique Democracy."

87 Filipinos embraced the opportunity; more than 200,000 students had enrolled in primary school by the end of 1902, and 20,000 had enrolled in secondary school. Difficulties in recruiting American educators overseas opened opportunities for Filipino teachers, who by this time filled three-quarters of 4,000 teaching positions. Abinales and Amoroso, *State and Society in the Philippines*, pp. 119–123.
88 Anderson, "Cacique Democracy."
89 Abinales and Amoroso, *State and Society in the Philippines*, p. 119; see also McCoy, *Policing America's Empire*, pp. 59–235.
90 Twain, "To the Person Sitting in Darkness," p. 170.
91 As quoted in Carl Schurz, "The Policy of Imperialism, Address by Hon. Carl Schurz at the Anti-Imperialist Conference in Chicago," October 17, 1899. Liberty Tracts 4 (1899) Chicago: American Anti-Imperialist League, p. 1.

2 Map

U.S. Colonial Science, Geo-Politics, and the Remapping of the Philippines

Atlas de Filipinas

The first U.S. Government atlas of the Philippines featured an odd imprimatur: two title pages (Figure 2.1). The first, in Spanish, and framed by a trellis of local flora, credits the work of unnamed Filipino draftsmen under the supervision of Father José Algué, Director of the Manila Observatory; the second, less ornate, lists the volume without authorship only as "Special Publication No. 3" of the U.S. Coast and Geodetic Survey.[1] The *Atlas de Filipinas/Atlas of the Philippine Islands*, containing 30 full-page color maps depicting distributions of peoples, active volcanoes, and earthquake frequency across the archipelago, among other phenomena, along with maps and landscape views of principal islands and regions, remains visually striking and impressive.[2] As the Head of the Philippine Commission, Jacob Schurmann, described the maps for U.S. Secretary of State John Hay, the high quality of mechanical execution "spoke for itself"; as for their accuracy, "we can only state that they fairly represent the present state of knowledge."[3] But while the "absence of accurate surveys of many of the islands" was, for Coast and Geodetic Survey Superintendent Henry Smith Pritchett, a "serious drawback," the Americans were evidently pleased with their acquisition.[4] Even before the December 1898 Treaty of Paris had rendered the archipelago from Spain for $20,000,000, a collaborative arrangement with Algué (and the Jesuit order) was in place to appropriate the *geography* of the Philippines for a price of $1085, plus 1,000 copies of the completed work; it was a fee, in Schurmann's estimation, which "border[ed] on the ridiculous."[5] The *Atlas*'s dual title pages were a reflection of this cartographic bargain, hinting at tensions in the nature and ownership of geographical knowledge during a moment of territorial transition.

Claiming authority, within the lines of the Paris Treaty, over some 1,000 islands (as were then estimated) and perhaps seven million new subjects, the new U.S. colonial administration quickly turned to practices of mapping—military, bureaucratic, and scientific—as key representational techniques in its efforts to transform the Philippines into knowable, governable, and exploitable territory. Indeed, the volume of cartographic production

DOI: 10.4324/9780429344350-3

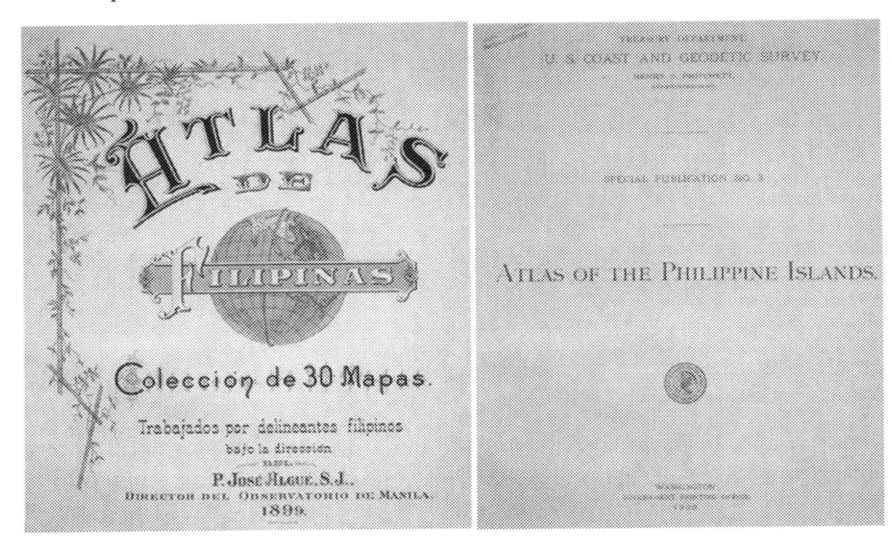

Figure 2.1 Dual title pages in the United States Coast and Geodetic Survey's (1900) *Atlas de Filipinas/Atlas of the Philippine Islands.*

under early U.S. rule, including both government and commercial publication, reflected a profound acceleration of mapping activity from Spanish colonial norms. As late as 1897, according to one Spanish archivist, just 140 maps, charts, and plans of the archipelago had been acquired during Spain's over three-century-long intervention; by 1903, the U.S. Library of Congress had collected 860.[6] And yet at this point, many of the most active map producers, from the Coast and Geodetic Survey to the Insular government's scientific bureaus to commercial cartographers like Rand McNally, boosted by the development of cheap wax engraving technology, were just beginning their extensive cartographic reproduction of the Philippines and "American Pacific."[7] Interestingly, for the Filipino historian Carlos Quirino, this explosion of mapping at the onset of the U.S. colonial period marked a fitting *endpoint* for his remarkable annotated history of Philippine cartography from 1320, in that, "starting with 1900 a plethora of maps has been issued, which would have added to the bulk of this book, without adding much to cartographical lore."[8] Conversely, this chapter turns precisely to the explosion of cartography under U.S. rule, and with the uses of scientific mapping in the making of colonial state space in the Philippines during the first decade of colonial governance. That today there are *seven*-thousand islands identified in the archipelago underscores the extent to which even the geography of the Philippines has been a work-in-progress. Under the Taft-Forbes regime, as I sketch in this chapter, scientific and cartographic practices of survey, representation, and inspection were integral not only to the more accurate representation of the islands on paper but also in linking governmental knowledge production with territorial transformation and the practical extension of U.S. colonial sovereignty.[9]

A range of historical scholars have effectively situated government mapping projects, alongside the development of surveys, censuses, and statistics, as part of an evolving, peculiarly modern form of state territoriality.[10] Maps were of special importance to the late colonial state, as the problem, which Foucault identified, of "connecting the political effectiveness of sovereignty to a spatial distribution" was one that was perceived acutely by officials and bureaucrats of the late nineteenth and early twentieth centuries.[11] As technologies for organizing and representing all manner of geo-coded information, maps and mapping practices were expressive of "the relations of power which pass via knowledge."[12] In the Philippines, these cartographic techniques, as they were employed in a quickly emerging (but initially quite limited) apparatus of government science, offered the Insular government a key model for organizing governmental knowledge production across diverse and unstable terrain.

Although the American delegation to Paris in 1898 had, in coming to cartographic terms with Spain, incorporated geodetic coordinates from British Admiralty charts, the acquisition of the Philippines would also prompt new practical and ideological relationships for the colonial state with the inheritance of Spanish geographical knowledge. In the next section, I explore the early efforts of an American scientific bureaucracy in the Philippines that, its leaders believed, would quickly supersede the accomplishments of Spanish colonial science, helping to contain the archipelago within a modern classificatory grid. Scientific surveys and mapping of the Philippines—including topography, geodesy, and ship-based mapping of coastlines; geology; ethnology; botany; and forestry—were extended across the archipelago, alongside major investments in tropical medicine and laboratory sciences.[13] In the territorial practices of these survey-and-mapping sciences, the geopolitical and the scientific were closely entwined, reflected in a cartographic model of governmental knowledge production that worked not only to produce more accurate representations of the islands on paper but also to facilitate the extension of territorial sovereignty, and governmental techniques of calculation, at multiple sites and spatial scales. The chapter continues by tracking the development of a distinctive geo-politics of knowledge (alongside biopolitical dimensions of colonial science and medicine) through the work of an array of geographical knowledge producers, from government scientists and bureaucrats to officers of the Philippine Constabulary, in their efforts to fix the position of Philippine lands, resources, and peoples within grids of power and knowledge.[14]

Insular Science and Geo-Politics

The establishment of a colonial scientific bureaucracy in Manila helped to define new forms of U.S. territoriality and governance under the Insular

Government. As early as 1900, even as the Philippine-American War was evolving into widespread guerilla warfare, the Philippine Commission created Bureaus of Forestry and Mines to promote exploitation of the archipelago's natural resources and to extend regulation over those domains.[15] Beyond laying bare a new terrain for exploitation, installing a scientific apparatus, largely under the administration of the abrasive Secretary of the Interior (and Philippine Commissioner) Dean C. Worcester, the Insular Government embraced a vision of a modern, scientific, and progressive American governance and colonial tutelage, supposedly a world apart from Spanish despotism.[16] In 1901, a Bureau of Government Laboratories was established to "perform all of the biological and chemical work of the government under the direction of one chief."[17] For Dr. Paul Freer, the University of Michigan chemist that Worcester had recruited to direct the Bureau, the business of uniting the government's laboratory work was thus precisely to "scientize" colonial policy, reflecting a departure from other tropical governments.[18] Mobilizing science would also provide the American colonial cadre with an exceptionalist response to both anti-imperialist and Filipino nationalist critiques of U.S. colonialism in the archipelago, one that simultaneously recovered and elided liberal and democratic impulses.

Historian Warwick Anderson is not wrong in observing that the Insular Government principally favored laboratory work in its scientific endeavors.[19] In *Colonial Pathologies*, his trenchant history of American colonial medicine and public health in early-twentieth century Philippines, Anderson traces the development of laboratory spaces (among other settings, including the hospital, sanitary public market, and leper colony) as key nodes in a range of distributed government practices that targeted hygiene, sanitation, and everyday life as objects of knowledge production and sites of government intervention. Biological laboratories were thus built "for investigation into, and scientific report upon, the causes, pathology and methods of diagnosing and combating the diseases of man and of domesticated animals" for the use of the bureaus and departments, while chemical laboratories would allow scientists to break down all manner of food, drug, plant, and mineral resources from across the archipelago.[20] Under progressive U.S. governance, the Board of Health and Bureaus of Agriculture, Forestry, Mines, Public Instruction, Public Works, and Customs Service would all require laboratory work, supporting everything from routine medical testing to chemical analysis and investigations of tropical flora and fauna, minerals and soils, and materials testing of coals, cements, and road materials.[21] In 1904, construction was completed for a large two-story laboratory building, comprised of separate biological and chemical wings with a modern scientific library in-between. The Government Laboratory building, stocked with cutting edge scientific equipment (compressed air, vacuum pumps, gas generators, microscopes, and microtomes for splitting biological specimens) and featuring its own power house and serum laboratory in the rear of the complex, provided a set of architectures and technologies designed to meet the challenges of late colonial governance, and thus to project U.S. power in the Philippines, through scientific means.

Located in the Ermita district of central Manila on the site of the former Spanish exposition grounds, the new laboratory offered U.S. colonials a model of control in an unruly tropical environment. It also served as a staging ground for a range of disciplinary practices that allowed public health scientists to re-imagine the entire archipelago as "a laboratory of hygiene and modernity."[22] Anderson recovers the workings of the colonial scientific laboratory as not only an *experimental* site, however, but also an *industrial* one, characterized by banal activities of processing, analysis, and manufacture of materials (such as the smallpox virus for use in serums), and the testing of human corporal matter at an immense scale. Staging an assault on "promiscuous defecation," among the problems associated with the persistence of water-borne diseases and parasites in Manila's water supply, depended on a vast production and circulation of materials, along with the formation of new subjectivities outside the laboratory gates. In this sense, the specialized site of the laboratory, for Anderson, is understood in its relations to socially and geographically distributed economies of public health, biomedical citizenship, and biopolitical interventions, reflected in both the local architectures of science and in the broader networks of exchange through which blood and fecal samples, and associated practices, regulations, and techniques, would travel. American colonial doctors and scientists, as Anderson observes, thus "hoped that with much time and effort the disorder and promiscuity of the islands might be subdued so that colonial space might come to resemble the controlled conditions of the modern laboratory. Yet this expansionist trajectory, in which the laboratory is imagined simply as a territorializing technology, can disguise a more complicated scalar politics." That is, if laboratory practices served as a means of extending government control over territory, Anderson suggests a more complex geographical relationship at work, wherein "it was equally important to make a *distinction* between colony and laboratory, if only to emphasize the superior culture of American modernity and how much more progress Filipinos and their country had yet to make. The flexible scale of the colonial laboratory—its capacity to magnify and diminish its focus—allowed a play of differentiation and assimilation."[23]

The expansion of the Bureau of Government Laboratories into a larger Bureau of Science, under Worcester's Interior Department, in 1905, incorporating the Bureaus of Mines and Ethnology alongside Divisions of Bacteriology and Medicine, Botanical and Zoological Sciences, and Chemistry, in some ways reinforced the central position of the laboratory sciences at the hub of scientific Manila.[24] But the move also reflected the wider ambitions of Worcester and the Philippine Commission, and the roles conceived for what might be called the Insular sciences, in the classification, evaluation, and governance of resources and territory outside the laboratory. Within a year of the laboratory's opening, Freer had called for the construction of two additional wings to accommodate scientists from the Bureaus of Forestry, Public Lands, Agriculture, Ethnology, and Mines, and to provide adequate space for visiting scientists. The Bureau's space needs would be alleviated instead by the purchase of the nearby Oriente Hotel for use by the Bureaus of Forestry and Agriculture, and by squeezing zoologists and botanists (and their increasingly bulky collections) into four rooms of the existing building and herbarium. But laboratory work in Manila was also a territorial and geographical project. Without pitting science in the field against the more fully professionalized laboratory sciences, then, it is useful to view the Insular state's problematic of knowledge production in the Philippines through the lens of a *cartographic* model of science—organized around survey, expedition, spatial data collection, map making, and the geographical mobility of observers—as a parallel apparatus alongside the laboratory, a territorially focused *geo-politics* of science, in other words, that articulated with Anderson's biopolitics among the operative discursive practices of the American Insular state. While, in historical terms, the (un-hyphenated) discourse of classical or imperialist geopolitics was developed chiefly around nation-states (along with "races" and "civilizations") acting at a planetary scale, the notion of geo-politics is used here to emphasize the ways that (colonial) state power was projected through the construction of geographical knowledges, perhaps helping to extend our understanding of biopolitics to what Thongchai Winichakul called the geo-body. For as Thongchai described the emerging nineteenth-century geo-body of Siam (Thailand), it was the map that "anticipated social reality, not vice versa. In other words, the map was a model for, rather than a model of, what it purported to represent."[25] In early twentieth century Philippines, as I sketch below for U.S. colonial projects of forestry, geodesy, geology, and ethnology, among the defining features of the new scientific bureaucracy were both the centrality of mapping practices and the presumed links between the construction of knowledge and the material transformation of the places, peoples, and things that the maps depicted.

Mapping the Field

The American (re)mapping of the Philippines was expressed in a range of intersecting discourses—military, commercial, scientific, and bureaucratic. The scientific project began modestly. One set of integrated "scientific surveys of the Philippine Islands" was contemplated, under proposed U.S. federal sponsorship, to have included a range of sciences organized around the mapping of terrain, vegetation, and the human landscape, including marine hydrography and geodesy, topography, geology, forestry, botany, zoology, and anthropology. Designed in broad strokes by a National Academy of Sciences committee in 1903 with the support of Secretary of War Taft, the proposal was introduced to Congress, perhaps belatedly, by President Roosevelt in 1905, calling for an extensive eight- to ten-year program of scientific surveys of the archipelago.[26] With a cost estimated at one million dollars (about 29 million today), it is not surprising that Roosevelt waited until after the 1904 Presidential election to bring the matter to Congress, going so far as to instruct the Philippine Commission *not* to include a planned endorsement of the program in its 1904 Annual Report.[27] But as Roosevelt would go on to suggest—in language prepared for him by Taft—in the wake of his 1904 election win, the islands of the Philippine archipelago presented "as many interesting and novel questions with respect to their ethnology, their fauna and flora, and their geology and mineral resources, as any region of the world."[28] The National Academy had considered "the desirability of instituting scientific explorations of the Philippine Islands," and specified that "scientific surveys which should be undertaken go far beyond any surveys or explorations which the Government of the Philippine Islands, however completely self-supporting, could be expected to make." Recommending the establishment of a U.S.-based governing board of the survey, Roosevelt emphasized the more-than-local value of the endeavor: "The surveys, while of course beneficial to the people of the Philippine Islands, should be undertaken as a national work for the information not merely of the people of the Philippine Islands, but of the people of this country and of the world."[29] Or, as Taft had emphasized for Roosevelt, the surveys should be "treated as a national duty in the interests of science."[30]

The claim that there *was* a national duty in the interests of science would itself have been quite novel in early twentieth century Washington, and not surprisingly, the proposed survey was not without controversy. Shortly after the National Academy committee study was initiated, the Philippine Commission—evidently blindsided—resolved that the "Government of the Philippine Islands has no objection to such a survey, but on the contrary welcomes it; but that, whether Congress takes this action or not, it expects in the course of a decade to institute many surveys on its own account, and it does not desire to be put in the attitude of requiring an appropriation from the United States Government for this purpose." The costs of the survey, Commissioners insisted, should not be tallied among the expenditures imposed on the United States from the "Government of the Philippines."[31] Freer expressed to Colonel Clarence Edwards, Chief of the War Department's Bureau of Insular Affairs, that he was "personally … very anxious to see the surveys carried out, as they are fundamental to our knowledge of the Islands," but noted that "it would seem to me that in consideration of the work which has already been done in the Philippines in a scientific way, and also in consideration of the fact that true cooperation can not be obtained without representation, the Government of the Philippines should be actually represented on the Governing Board of these surveys." Beyond the governing board, Freer reminded Edwards, the entire project would depend on cooperation among "the men who have been for years in the Philippines have the knowledge necessary to assist in the surveys to the fullest extent; they know the territory, the needs of the

Islands, and the means by which results can be accomplished in the quickest way."[32] Later, Harvard geographer William Morris Davis voiced his own concerns over the "Washington centralization" of the proposed surveys, asking Edwards, after a conversation with Cameron Forbes during the latter's stateside visit, why Secretary Taft should not "secure the control of the proposed surveys to the Philippine commission, instead of placing control in hands of Washington bureau chiefs?"[33] The National Academy's plan for integrated scientific exploration and survey of the Philippines, though endorsed by Roosevelt, would go unrealized, and yet the "Scientific Surveys of the Philippine Islands" were in some ways pursued, in *ad hoc* fashion, as projects of the Insular scientific bureaus, along with the U.S. Coast and Geodetic Survey, which would continue to make fundamental contributions to geodetic knowledge and cartography of the archipelago from the outside-in, that is, from shipboard perspectives.[34] Forestry would be pursued, initially, from a similar subject position.

Of course it was one thing to propose (or bicker about) scientific surveys, and another to carry them out. The first director of the Forestry Bureau, Captain George Ahern, noted smugly that Spain had not established its forestry department until 1863, "some three hundred and forty years after their occupation of the islands" had begun, but observed that the "forest zones had not been surveyed and reserved, as the last Spanish land law of 1893 had contemplated."[35] Given that Spain's small colonial garrison had, during the 1896–1898 Filipino revolution, lost its capacity to project power across much of the archipelago, Ahern's expectations seem a bit out of touch. But his comments (and early work at the bureau) are telling precisely in the normative spatial categories that they produced, emphasizing the *yet-to-be surveyed*. Still, the new chief bemoaned in his initial report that the Bureau had been beset with practical impediments, from difficulties in identifying the large variety of tropical tree species in Philippine forests to filling its labor needs to gaining access to distant forests amid persistent conflict. Ahern's vision of Manila as the lumber emporium of the Far East, serving markets in Hong Kong, Shanghai, Singapore, Sydney, and Nagasaki,[36] would require not only adequate staffing but also the winning of the peace in regions remote from Manila. And however comfortable the premises at the old Oriente Hotel, foresters, among other colonial survey scientists, encountered significant obstacles to executing their work in the field, let alone in realizing any neat visions of scientific governance.

While Spain had employed, at last count, 66 foresters, 64 rangers, and an additional staff of 40, the Americans would initially struggle to maintain a staff under 50, including Ahern, one inspector, one botanist, and ten assistant foresters.[37] As Ahern complained, after foresters had begun to learn the varieties of dyewoods, gums, resins, and other tropical species, and trained in "instructing ignorant native loggers in the principal requirements of the forestry regulations," they were then easily poached by lumber companies offering salaries above the government pay grade. The violence of the ongoing conflict had also presented serious challenges to staffing, as "a disposition was shown by the native officials to avoid service beyond Manila." Even in the Manila suburbs, Ahern reported, "at times the native officials would receive threatening notices, and as quite a number of natives friendly to Americans had been captured and murdered by the insurgents," many officials were unwilling to expose themselves in the work of inspecting timber rafts.[38] Two Filipino rangers had disappeared on the job, one of whom returned a month later, claiming to have bought his freedom from *insurrectos*, while the other had not been heard from. If the Philippines represented, as Ahern insisted, "a vast virgin field for scientific investigation," then along with its essential work of industrial regulation, that science existed with a specter of violence, and much of the archipelago's extensive upland forests remained inaccessible.[39]

In October 1902, the American chief forester Gifford Pinchot visited the Philippines to assist in working up a plan for both exploiting and conserving the archipelago's forests. Three years before Roosevelt appointed him founding director of the U.S. Forest Service in Washington, Pinchot spent six weeks touring the archipelago, doing most of his forestry, as historian Greg Bankoff notes, from the deck of a ship.[40] Enjoying the pleasures of good food and an "unusually comfortable" state room, Pinchot cruised 2,300 miles around the archipelago on-board the gubernatorial yacht, gazing through field glasses at forested (and deforested) coastal inlets and slopes.[41] Making occasional incursions to insular interiors to more closely examine the tropical tree species that were, by his own admission, entirely new to him, Pinchot also evaluated the conditions of contemporary logging in some locations and contemplated evidence of deforestation in the already barren slopes of Panay and the Ilocos coastline in northern Luzon, in contrast with largely intact forests of Mindanao in the southern archipelago. Insisting, with Ahern, that Americans would manage the Philippine forests more efficiently than had the Spanish, Pinchot estimated, nonetheless, that half of the archipelago's forests had already been felled, departing from the Bureau's rosier estimates of 65–70% of forests still standing.[42] Pinchot's incursions to the interior uplands, under escort of six soldiers, brought him into closer contact with unknown tropical tree and plant species and what he saw as exotic peoples and places, but also revealed the desperate conditions of social life in many of the places he visited. Six months into a cholera pandemic that would ravage the Philippine countryside for two years, and occurring alongside a brutal counter-insurgency which, along with the rinderpest, was decimating Philippine farms and livestock in targeted regions, small wonder that the forests had become, in Pinchot's estimation, a "refuge of the disaffected."[43] But Ahern's Forestry Bureau, operating with a freedom of experimentation and intervention not possible in incorporated U.S. territories, sought to expand the purview of Philippine forestry through the organization of surveys, reservations, and "scientific management" plans in which Philippine forests could be viewed as proving grounds for American scientific forestry.[44]

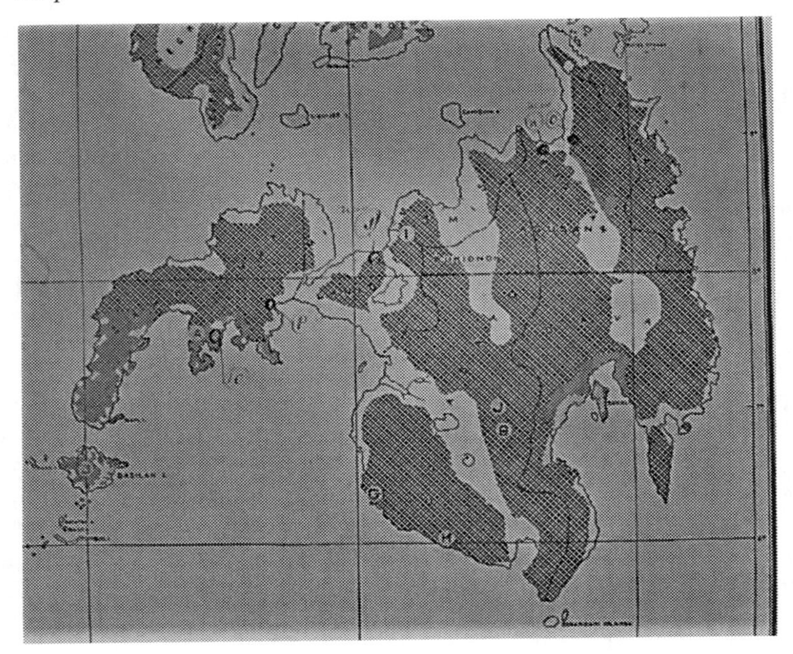

Figure 2.2 Detail of Mindanao and vicinity from George Ahern's "Forest Map of the Philippines," Bureau of Forestry (1910), NARA Record Group 330/21/18/5-1, Cartographic and Architectural Section. The dominant cross-hatching, richly pigmented in green, depicts the category "unexplored commercial forest."

The Bureau's (1910) "Forest Map of the Philippines," reflecting the expansion of field surveys during 1908–1910, enveloped the archipelago in a seven-fold characterization of forest type and land use categories, drawing covetous eyes to the green, hash-marked category "unexplored commercial forest" blanketing Mindanao, Palawan, and northeastern Luzon, among other settings (see Figure 2.2).[45] With Ahern convinced that proper regulation of forests depended on comprehensive resource mapping, "lines of rangers under the guidance of trained foresters literally walked the length and breadth of the archipelago's forests, identifying the species and noting down the size of the trees they encountered."[46] By 1914, timber was second in value only to agriculture among Philippine exports, reflected in a correspondingly rapid depletion of the archipelago's forest cover, and a Forestry Bureau which purported to "fund itself."[47]

Close connections between state projects of knowledge production and the cartographic construction of space, territory, and resources also existed in the more technically sophisticated mapping sciences. Geodesy, the science of determining the precise figure of the earth, was prized, amid the pervasive naval militarism of the era, as an emblematically modern, practical science. Executed with its greatest precision as an off-shore, ship-based activity, geodetic work was not placed under the offices of the Insular government. Rather, the U.S. Coast and Geodetic Survey—which had already put its *imprimatur* on Algué's *Philippine Atlas*—was called over from work in the Eastern Pacific and engaged to situate Philippine coastlines on the modern geodetic grid.[48] Though delayed by events of the Philippine-American War, the Survey quickly established points for determining latitude and longitude and extending subsequent triangulation surveys that, in turn, provided the basis for a range of cartographic work in the Philippines, including land-based topographic surveys and resource maps, cadastral maps, maps featuring public lands and mining claims, disease mapping, and base maps for all manner of geological, biological, ethnological, and other geo-referenced scientific studies.

By 1910, according to the Coast and Geodetic Survey's "sketch of general progress" (see Figure 2.3), most of the Philippine insular coastlines and coastal zones, and the spaces between them, had been surveyed, producing an integrated marine and terrestrial space that reflected both commercial and strategic priorities.[49] The progress map, which showed where triangulation, topography, and hydrography surveys had been executed, points for latitude and longitude determined, tidal and magnetic observations made, and deep sea soundings taken, was not itself a scientific map of terrain, though its coastlines were presumably geodetically accurate. It was rather a map of mapping itself, a geography of scientific practice, but one which also betrayed a certain cartographic anxiety. Such cumulative, incrementally-produced mappings, which included "progressive military maps," Signal Corps and Philippine Constabulary "progress maps," and Coast and Geodetic Survey general progress sketches, were employed widely by colonial agents and institutions that "wished to display progress in a graphic manner," as Col. Clarence Edwards, chief of the War Department's Bureau of Insular Affairs, put it in an early missive to Governor Taft requesting additional maps and geographical information from Taft's bureau chiefs and provincial supervisors.[50] Cartographic anxiety, in such mappings, was expressed not as a crisis of representation but one of imperial governance, reflecting a desire for filling in the map, and representing the unknown as rationally governable territory, which envisioned a world of colonial problems that the completion (or perfection) of the map could resolve.[51]

Examining the military gaze in particular through the construction of a (roughly concurrent) progressive military map of Puerto Rico, sociologist Lanny Thompson describes "a way of seeing objects and of speaking about them, a field of visibility and a mode of enunciation" in which practices of mapping served an imperial state apparatus that "created knowledge, established relations of power, and defined subjects."[52] An explicitly cartographic model of knowledge construction (and governmental rationality) similarly engendered a range of contributions, both practical and ideological, for the Philippine Insular Government, offering technologies of wayfinding, information storage, and communication which, while relying on the

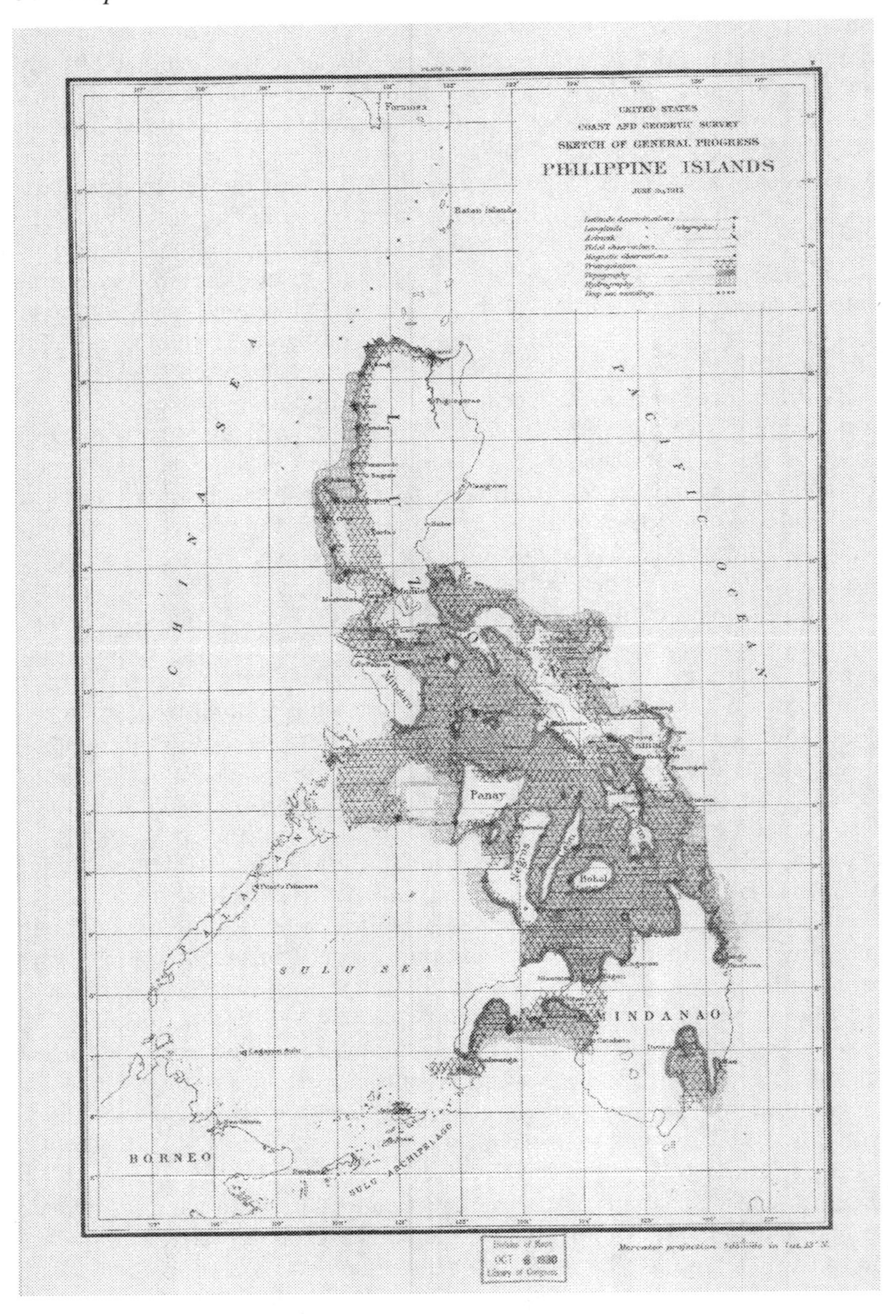

Figure 2.3 United States Coast and Geodetic Survey (1912), "Sketch of General Progress, Philippine Islands, June 30, 1912."

Source: Courtesy, Geography and Map Division, Library of Congress.

geographical fallacy of maps as value-free representation, served to render territory and resources as actionable fields. It was clearly no co-incidence that the Bureau of Science's relatively sparse 1907 "Map Showing Principal Mineral Districts" across the archipelago, featuring known deposits of gold, copper, iron, and coal, also included existing railroads and those "partly completed or projected."[53] Such juxtapositions, which we can imagine appeared as perfectly natural to the map's readers as well as its producers, offer traces of the meaning-making work of maps as agents of geographical transformation. For as Thompson argues in the Puerto Rican context, the practices of colonial mapping, linking science, governance, and geographical knowledges, reflected cartography as "an instrument of appropriation, reconfiguration, and modification, not only of landscapes but of colonial subjects and populations."[54] The mapping of human difference, to which we turn below, was another important means of grappling with the field in the construction of new and actionable geographical knowledges.

Race, Non-Christians, and the Human Sciences

Ideas of "race" and reproduction of racial categories served as key tenets of U.S. imperial rule in the Philippines, offering Insular officials, government scientists, and colonial apologists both a descriptive lens onto, and putative explanation of, difference among peoples. Some categories were positively appropriated from Spanish colonial precedents, notably the peculiar binary of the *non-Christian*, explicitly territorialized in the Special Provinces, which encompassed both the animist mountain peoples of northern Luzon and the "Mohammedan" Moros of the southern archipelago. But American colonials found that reproducing these categories—while territorializing them as distinctly *non*-democratic settings for paternalist imperialism in the Mountain and Moro Provinces (under the administration of the Philippine Interior Department and U.S. Army, respectively)—required a great deal of ideological work. Alongside statistics, the human sciences thus contributed to this work, the project of a colonial state which, as Kramer has argued, "organized itself around new forms of knowledge-production, including the generation of novel racial formations."[55]

Most prominently, the Philippine census, launched in 1903 (and published in 1905), would employ more than 7,500 Filipino census takers to canvas the islands, mapping the "structure of racial difference," as Vicente Rafael describes it, while establishing "the privilege of a particular race to determine the borders of those differences."[56] In regions where insurgent and counter-insurgent warfare persisted, the Constabulary sometimes intervened to tamp down resistance enough to allow the census to take place. Meanwhile, the census project helped to produce new relationships for the Philippine Commission with local elite as sources of demographic information, paving the way for the expansion of limited electoral democracy in the form of the Philippine Assembly in 1907.[57] But the census was by no means the only such instrument as the Insular Government tasked itself with spatialized knowledge production in its (re)construction of a distinctively racialized geo-body. Like the paradox embodied in the American "empire of no empire" (and integral to that paradox), the contradictions of racial imperialist ideology could be difficult ones to hold in place: in order to justify the worthiness of the colonial endeavor and the competency of their own administration, Insular officials were compelled to demonstrate Filipino capacities for progress while, at the same time, emphasizing the racial, cultural, and historical limits of Filipino capacities for modern nationhood and self-rule that would require U.S. retention for "several generations," as Taft-Forbes regime boosters consistently advocated.[58] Preceding and alongside the census, ethnological mapping of "non-Christian" peoples in the Philippines, including investigations of the territoriality, resource usage, and usufruct rights, offered colonial officials a means through which the production of difference could be refashioned and linked to the administrative construction of spaces and territories.

Challenges of the field were no less daunting for the human sciences than other areas of survey-oriented research, and the links between scientific practices and issues of territorial and resource access were nearly as explicit. In the Fall of 1901, a Bureau of Non-Christian Tribes was established by the Philippine Commission, reproducing the category as an umbrella to cover the ethnology of diverse mountain peoples of the Cordillera and other island interiors along with the Moro peoples who lived mainly in the southern islands of Mindanao and the Sulu archipelago. Later renamed the Ethnological Survey under the Bureau of Science (and later the Division of Ethnology), it was modelled in some ways after the Smithsonian Institution's Bureau of American Ethnology in Washington, which had focused both scientific and advisory efforts on the status of American Indians in the North American West, but the geo-politics of expertise were, in the Philippines, even closer to the surface.[59] The Bureau of Non-Christian Tribes and its successors were charged with intersecting practical and intellectual objectives: "First, to investigate the actual condition of these pagan and Mohammedan tribes, and to recommend legislation for their civil government; and second, to conduct scientific investigations in the ethnology of the Philippines."[60] But as bureau chief David Barrows noted in his first annual report, Spanish "explorations and conquest in the Philippines, conducted with the utmost rigor for the first few decades after the arrival of Legazpi [in 1565], almost entirely ceased at the end of fifty years," leaving mountainous interiors and large parts of the southern archipelago "still unexplored and only imperfectly subject to governmental authority."[61] In turn, for Barrows, whose doctorate in anthropology (from the University of Chicago) was based on ethnological fieldwork in the North American West, the problem of enacting a government ethnology was understood in explicit relation to the territorial expansion of U.S. authority.

Barrows anticipated that much of his own efforts and that of senior staff "would be in the field, and there is scarcely a mountain or an island in the entire archipelago that does not invite investigation."[62] Formalized as the "Ethnological and Geographical Survey," Barrows's team of anthropologists, consisting of Barrows himself, assistant bureau chief Dr. Albert Jenks, government photographer Charles Martin, and, unnamed, an interpreter of Igorot and Ilocano dialects, American packer, and "native cook," would venture eastward from Baguio across the Cordillera Central in northern Luzon to Bayombong, Nueva Vizcaya, visiting *rancherías*—as the Spanish had dubbed the *indio* hamlets that constituted the primary mode of settlement in the interior highlands—along the way, accompanied in part by the provincial governor of Benguet.[63] "This intervening country is quite unmapped," Barrows enthused, "and a rough topographical survey of it will be made en route."[64] Equipped with waterproof tents, riding and pack saddles, and photographic and anthropometric instruments and supplies, Barrows initiated side excursions into such unmapped regions to visit "tribal" groupings that included Ifugaos, Italones, Ibulus, and Moyao. "The special object of touching these tribes," for Barrows, was "to determine their relationship to the great Igorrote family of the Cordillera Central," and, not least, to identify the "geographical habitat of the Igorrote and the number of their tribal and linguistic groups … and tribes bordering the Igorrotes on the north and east."[65]

Like other survey sciences, ethnology in the Philippines existed alongside (and depended on) a range of state-building projects, including the re-organization of civil governments in pacified localities, compulsory labor for road and trail construction, particularly in the Special Provinces (see Chapter four), establishment of schools under the Department of Public Instruction, and garrisoning of the Philippine Constabulary.[66] The Bureau would draw on these institutional resources in efforts to enact and expand its survey, seeking volunteer field workers from a range of colonial cadres, including both civil and military colonial staffs. Operating along what geographer Felix Driver has described, in a different context, as the "unsettled frontier" between emerging social science traditions and the adventurous colonialisms still present within them, Barrows adopted for his Bureau established norms of amateur data collection in botany, geology, and natural history, and other field-based sciences.[67] As early as 1901, Barrows casted a wide net among those capable of putting eyes on the local people, circulating a "special invitation" to "officers of the US Army and Navy in the Philippines, to inspectors of the Insular Constabulary, superintendents and teachers of the Department of Public Instruction, officials of the provincial governments, and to persons who through residence or investments in the more remote portions of the Islands have become familiar with the conditions there prevailing."[68] The special invitation served as both recruitment tool and rudimentary ethnological methodology, comprised of a set of "Instructions for Volunteer Field Workers" containing ten categories of suggested observations, each including a sub-set of more detailed questions.

What did Barrows want to know about the "more remote portions of the Islands"? Ethnological categories included in the Instructions were diverse, including names of the "tribe" as known to themselves and external communities; the "habitat or territory occupied" ("If possible get the native name for each "ranchería," "sitio," or village, and make a sketch map locating each, with notes as to hills, streams, and trails. This information is particularly valuable to the Bureau"); physical anthropology (cranial and other bodily measurements were encouraged if possible, noting that "physical data of much value can be gained by careful observation"); artificial deformations (tattooing, teeth filing, etc.); hair styles; dress and ornaments; religious or ceremonial attire; religious belief and customs; social organization; daily life and industry; and vocabulary (a list of 50 words to obtain was provided).[69] Following these "data of a scientific nature," a second set of questions turned more explicitly to practical matters: political, territorial, and security issues; potential for barter or exchange; agricultural capacities; whether "the country they occupy" was likely to attract settlers or prospectors; health and disease concerns; suggestions for "practical means through which the Government could improve their condition"; and a list of collectable items desired for museum display.[70] The Instructions thus provided a relatively brief but flexible survey instrument, reflecting an effort to discipline a new class of data collectors around the integrated scientific and political mission of the Bureau. Some questions, however, such as *whether* the peoples and the places they inhabited were to be made available for political calculation, had effectively been answered before they were asked.

By 1908, the Division of Ethnology's "Directions for Ethnographic Observations and Collections" had become considerably more elaborate, divided into 26 categories of observation, from "Geography and statistics" to material arts, dwellings, livelihoods, political relations, language, cosmology, and physical anthropology, with each category containing some 30 to 50 lines of questioning.[71] In the same year, the establishment of Mountain Province out of the Special Provinces of northern Luzon, administered under Worcester's Interior Department (and based largely on Worcester's own ethno-geographical divisions), can be seen as an attempt to further insulate governance of "non-Christians" from the influence of the new Philippine Assembly. Meanwhile the territorial addition, in the new map of Mountain Province, of a new corridor to the South China Sea, as discussed in Chapter three below, reflected additional geopolitical motivations.

Ethnological investigations, which included the accumulation of an unprecedented cache of government-produced photographic and cinematic imagery, contributed to a range of work products, including scientific publications, sometimes lavishly illustrated annual reports of the Insular bureaucracy, and publications—like *National Geographic*—targeted at popular audiences across the Pacific.[72] These data, images, and narratives, circulating in a wider economy of information linking race, empire, "savagery," and American exceptionalism, disproportionately depicted Filipinos as head hunters and "wild people" incapable of self-governance.[73] As the anti-imperialist critic (and former Insular circuit judge in the Philippines) James Blount was keen to point out, the intended effect of this broadcast of exotic images—what Blount called the "Worcester Kodak"—was to obscure more accurate understandings of Filipinos. Instead, the Worcester Kodak painted the Philippines with the broad brush of the islands' diversity, reinforcing stereotypes of a people unready for self-governance.[74] Ethnological maps, such as those depicting the "distribution of civilized and wild peoples," and commercial maps matching the location of peoples with the distribution of natural resources, contributed to this discourse by extending a special geographical fallacy characterizing disproportionately large territories as "wild," in contrast with the more densely populated, "civilized" lowlands which constituted about 85% of the archipelago's population.[75]

The Bureau of Insular Affairs' 1904 "Map of the Philippines" builds such categorizations into a color-coded geography of "Christian Provinces," "Moro or Mohammedan Provinces," and "Other Non-Christian Provinces."[76] Aside from vastly oversimplifying the population of the archipelago into these categories—a division apparently so self-evident that, despite being the map's dominant visual element, the subject did not need to be identified in the title—the map also performed the peculiar cartographic work of rendering each group as absolute within a given province, rather than reflecting a more nuanced and relational geography of settlement and identity. The scientific and governmental discourses of ethnology in the Philippines were thus classically imperialist and geopolitical, offering resources that could be used, as Babette Resurrección has argued, both to deny nationhood to the Philippines and to justify forms of U.S. imperialism in particular settings.[77] Colonial ethnology also contributed to the geo-political scientific project, as characterized in this chapter, of constructing, representing, and communicating knowledges of the geo-body, geographically coding the Philippines in ways that allowed the *bundok* to be reimagined as a site of colonial governmental intervention. Linked with geography, ethnology in the Cordillera lent cover for the creation (and reproduction) of Mountain Province as a formal territory coveted by retentionist colonial elements. The establishment of Mountain Province would also help to solidify Worcester's own authority over the region under the auspices of the Interior Department, as we explore further below. For the animist peoples of northern Luzon, however, the outcomes of such encounters could be complex. On the one hand, being mapped *into* these modest productions of knowable territory may have had profound immediate implications—for example, facilitating Insular state demands for taxes, including conscripted labor for roadwork, or reshaping the localized politics of patronage. On the other hand, being left *off* the map would also become consequential for some upland peoples who in turn struggled to access the benefits of ethnic membership in the modern state. As Resurrección notes, some individuals and "were actually fluid in their movements," belonging to multiple social groups, reflecting complex situations wherein they "acquired identities contingent to particular circumstances. By naming particular groups and partitioning their territory into officially recognized administrative units, the colonial government firmed up the ethnic boundaries of identity and habitat. These became reified and have proven enduring over time."[78]

In addition to its value in making colonial lands and resources visible for speculation, investment, and exploitation, the diverse knowledges produced under a cartographic and geo-political model of colonial science also provided currency for the (traditional) geopolitical discourses of colonial retentionists, including those who sought to establish enduring American enclaves in the Special Provinces even in the event of Philippine independence. That such visions would go unrealized, as we will see in the next chapter, did not make them less tempting to powerful agents engaged in the imperial moment. New ethno-geographical knowledges offered resources for an enterprising strain of American geopolitics which sought, through both formal and informal means, to hive off resource rich territories from the control of "Hispanicized" lowland Filipinos, meanwhile establishing paternalist colonial relations with idealized non-Christian subjects. Perhaps such relations could, in turn, be used to justify a continued colonial presence. More immediately, however, Barrows's ethno-geographical survey provided the Insular state with the promise of extending and deepening sovereign territorial engagements while also providing currency in debates over U.S. Philippines policy, Pacific geopolitics, and the nature of twentieth-century U.S. imperialism.

A Cartographic Colonial State

> The distinction between the material world and its representation is not something we can take as a starting point.[79]

The production of geographical knowledge under the American colonial state was by no means limited to the work of scientific experts. The cartographic reconstruction of the Philippines as a U.S. Insular territory also depended on a host of mundane local mapping practices which, though far from geodetically precise, constituted the frontiers of geographic knowledge for the Insular government and the Army at different moments and places, and reflecting the everyday survey and mapping practices through which the state sought to produce *working* geographical knowledges of the territories it inhabited and intended to govern. One "progressive military map of the Philippine Islands," for example, was launched under the War Department in 1905 by initiating a series of "field skeleton" maps, based on the best maps available, which were furnished to officers assigned to verify (ground truth) spatial data, mark corrections, and add new information on wide ranging matters of geo-location relevant for military purposes, including altitude, location of high points, ridges, wooded areas, streams and ferries, and to regularly circulate their findings—to be incorporated into subsequent editions of the progressive map—through updating and annual reporting functions.[80] Here too, the opposition between the material world and its representation, as the epigraph suggests, was a product of socio-political practices, that is, the *representational* work of occupation, colonial administration, and counter-insurgency, rather than a starting point for depicting an actual world distinct from those practices. Or, as historian of cartography Brian Harley has captured the fullness of this paradox, "The map is *not* the territory, yet it *is* the territory … cartography is part of the process by which territory becomes."[81]

The Philippine Constabulary was established in 1901 on the model of the Spanish *Guardia Civil*, which had served similar purposes. U.S. armed forces in the Philippines were reduced to 72,000 in 1905 and 13,000 by 1910. Alongside this drawing down of U.S. troops, the Constabulary, a force of about 6,000 Filipinos officered mainly by 230 Americans, many detailed from (or upon mustering out of) the U.S. Army, was increasingly leaned on to "keep the peace" amid ongoing conditions of political and structural violence—violence that was systematically downgraded to matters of brigandage, banditry, and corrupt caciquism by an Insular state anxious to put the war behind it. Its work included investigation and adjudication of local disputes (outside of Manila), providing armed escorts for government expeditions, such as Interior Secretary Worcester's annual inspection tours of the Special Provinces and visits of provincial and sub-provincial lieutenant-governors; the suppression of brigandage and insurrection; and sometimes, taking sides and settling scores in brutal local engagements or engendering such conflicts.[82] In one such incident in 1911, recounted approvingly by Worcester for Governor-General Forbes, a Constabulary expedition aimed at securing unlawful firearms from "recalcitrant settlements" in the Bontoc and Kalingas sub-provinces had set out, under Lieutenant-Governor Walter Franklin Hale, deliberately "to punish a lot of lawless savages who had recently been engaged in murder and rapine, and who had persistently refused to yield to the authority of the established Government, and it did so."[83] The result, Worcester insisted, had been the establishment of peace in the region and the increasing recognition of U.S. authority. Hale's expedition had included an army of 600 Kalinga warriors from surrounding villages and rancherías. "Some shacks (the word 'houses' is really not applicable) were burned," Worcester admits, "I approve of this action. So far as possible pigs and chickens were killed. I approve of this action." He also observes, but does not sanction, more grisly consequences, as when a Kalinga woman and child were killed by a Constabulary soldier who, following local custom, claimed the woman's jawbone. Worcester confessed only that it was, "impossible to make an omelet without breaking eggs."[84] Despite internal contests over its proper use, the Constabulary, operating with a relatively small force across extensive insular spaces and a challenging, mountainous terrain, would garner a reputation for effectiveness among the Taft-Forbes regime. Certainly, it was a source of stability for the Insular state, providing opportunities for patronage, income, and social mobility for those who had thrown their lot in with the Americans. Garrisoned across the archipelago, the Constabulary also extended its participation in state-building and sovereignty-extending exercises through active mapping and archival practices that could be incorporated into the duties of its officers at the district-level.

"Progress mapping" was one of many discourses contributing to the U.S. remapping of the Philippines, but it was one which, like the scientific (and quasi-scientific) geo-political endeavors sketched above, placed high value on the production and circulation of new geographical knowledges, particularly geo-coded information that merged ethnological and physiographic elements in the construction of spaces, boundaries, territories, and "habitats." The Philippine Constabulary's model of the progress map, like the progressive military maps it was modelled on, thus served as a cumulative geographical archive, carried out at the district level, to which inspectors in the field were instructed to contribute during the course of their duties, circulating their observations up the chain of authority via blueprints, tracings, and sketch maps.[85] The progress maps were thus expected to be of practical value for Constabulary patrols, escorts, and district offices charged with policing and peacekeeping duties across areas that were characterized by the persistence of blood feuds as well as counter-insurgent warfare.[86]

New cartographic demands were codified in a 1906 directive from the Acting Director of the Constabulary to the District offices in which each senior inspector was instructed to divide his district into sections of not more than 400 square miles, and each station master required to "keep a progress map" at a large scale of two inches per one mile "of each of the sections in which his command operates."[87] Along with a compass, each senior inspector was to be furnished with a T-square, a quantity of drawing and tracing paper, a protractor, and a right-angled triangle. "When practicable," senior inspectors were instructed to "habitually carry the compass when on expedition, and from vantage points on roads and trails, and on prominent hills, sites, peaks, mountains, rivers, bends, rocks, church towers, etc., and record same for purpose of location by triangulation and plotting on progress maps." Maps should include the names of mountains and hills, slopes, peaks, and rivers, as well as rancherías, pueblos, and suitable locations for making camp (see Figure 2.4). After expeditions, new points obtained were to be entered in ink on the progress map, and inspectors instructed to render a new tracing that could be forwarded with a report to the senior inspector who would, "without delay, plot on the progress map kept by him the additional points and information and forward the tracing to the

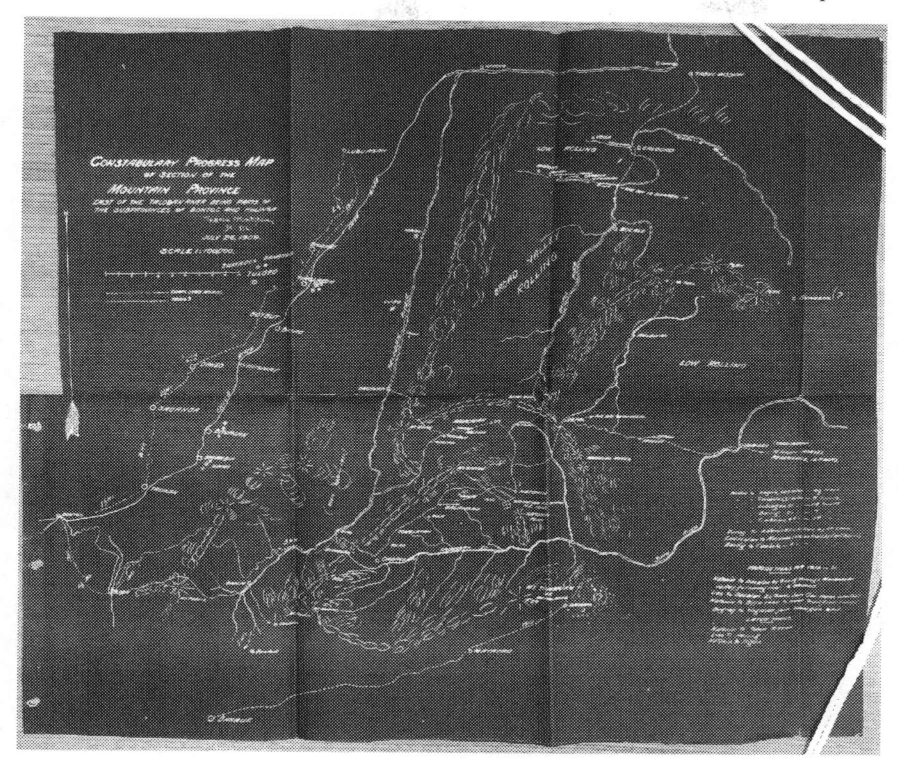

Figure 2.4 Philippine Constabulary progress map from the Bontoc and Kalingas sub-provinces, Mountain Province, in white on blue background, July 26, 1909. Worcester Philippine Collection, Worcester Papers Vol. 15.

Source: Courtesy, University of Michigan Library (Special Collections Research Center).

District director."[88] Sketches of routes traveled were also to be included in expedition reports, some of which survive as works in progress, blueprints scrawled over in dark pen.

While new geographic knowledges thus circulated "upwards" institutionally via the tracings, working progress maps in the district offices, in which hand-drawn lines and symbols overlay the printed maps, served as the best available, and held obvious value to the station-master and other officers. Hence, it is interesting that the 1906 directive specifies that the progress maps must be transferred and receipted as parts of provincial and station records to succeeding officers—it was stipulated that the maps belonged to the station, not to the officer in charge. The need to *clarify* this rule is suggestive, highlighting another tension in the nature and ownership of geographical knowledge, for it implies that just the opposite—the map seen as a matter of personal rather than institutional authority—may have been occurring. Given the centrality of local geographical knowledge to their work (and to the advancement of their careers), a personal, proprietary sense of the progress maps is hardly surprising, and indeed calls attention to the intimate, situated, and positional nature of map construction and use. Constabulary officers had been enlisted as seeing eyes at the scale of experience, embodied agents of the Insular state whose maps would reflect not (yet) the absolute spaces of the geodetic grid but the paths of their travels in a relational, less charted terrain. The surviving maps, with their light etchings dominated by dark blue backgrounds, today appear less the instrument of a dominant colonial state than a record of grasping in the dark. And yet the maps retain traces of hegemonic practice, reflecting precisely the positionality of the cartographer in the landscape. The geographical knowledge produced was thus both dynamic and relational; and it was precisely in the maps' secondary and tertiary circuits, that they acquired their principal value and power, each trip, ideally, paving the way for more knowledgeable return travels.[89]

The Constabulary's progressive cartographic project perhaps came to its fruition in the distribution of a pocket-sized handbook, *Maps of the Philippines by Provinces*, which included large-scale sub-provincial maps along with a compendium of regionally organized information in a convenient portable format. It was, as Acting Director J.G. Harbord described it, in gifting the volume to Governor-General Forbes in March 1912, "a little book we use in the Constabulary," providing "information about the provinces, roads, trails, Constabulary stations, etc."[90] Alongside rudimentary maps, the volume included a range of geographical data in typescript at the sub-provincial level (over a standardized list of categories for each region), including statistics of population, area, principal towns, key agricultural, and mineral products, names of government and Constabulary officials, location of telegraph, and telephone offices and Army and Philippine Scout garrisons, narrative descriptions of trails, and Constabulary transportation resources—in Benguet, for example, this amounted to just three "native horses," one mule, and one carromata, a two-wheeled cart. The little book thus provided a practical micro-geography of the sub-provinces, based on data "as accurate as the means at our command can make it," as Harbord put it, "We find it convenient for quick reference on trips of inspection through the Archipelago."[91] Receiving the volume, with space on its pages for categories, left blank or filled-in, such as "Ladrone bands, or noted criminals, at large" and arms in the hands of police or privately licensed, Forbes, for his part, may have read a validation—in the book's air of uniformity and routine—of Taft's vision for the gradual but uneven transition in the American colonial Philippines from the suppression of insurrection to the banal policing of brigandage.[92]

Harbord's geographical gift had not been made to assist Forbes's wayfinding in the *bundok*, however, but that it "might be of use to you while in the United States," on the Governor-General's spring 1912 stateside leave. The Acting Secretary thus wished Forbes, when visiting the War Department in Washington, to ensure the continued detail of several Constabulary

Inspectors from the Army, amid grumblings that 700 officers had been "absent from their regiments" while assigned to the Constabulary.[93] As Harbord saw it, "These are men performing a work for the Insular Government that we are not prepared to meet by the detail of our own officers and they constitute between the Federal authorities and the Islands a portion of a very small number of the Army who are now in touch with the topography and people of this Archipelago. The intimate knowledge which the Army had of the Islands during the insurrection has, for practical purposes, disappeared."[94] Harbord insisted on the dynamic and embodied nature of the Constabulary's geographical knowledges, as reflected in the pocket atlas itself. Not confined to a separate, stable sphere of representation, as if such a thing were possible, the everyday work of a cartographic colonial state was rather "part of the process by which territory becomes."[95]

***

"Once power can be analyzed in terms of region, domain, implantation, displacement, transposition," Foucault maintained, "one is able to capture the process by which knowledge functions as a form of power and disseminates the effects of power. There is an administration of knowledge, a politics of knowledge, relations of power which pass via knowledge and which, if one tries to transcribe them, lead one to consider forms of domination designated by such notions as field, region and territory."[96] Or Insular Territory. If, as argued in the previous chapter, the category of the insular, borrowed from physical geography and the proto-geopolitical discourses of U.S. naval strategy, provided a geographical stand-in for the more explicitly political terminology of colony and empire, then the purpose of this chapter has been to trace some of the "relations of power which pass via knowledge" as they took form cartographically around the emerging U.S. colonial state in the Philippines, illustrating the close integration of mapping practices with the production of territory and new state spaces in an unsettled political landscape. Through a rough sketch of a range of institutionalized cartographic practices, rather than a fine-grained analysis of a single discipline, survey, or scientist, the chapter suggests that the mapping impulse was widespread in early state projects of ethnology, forestry, geodesy, geology, and other survey sciences, as well as paramilitary and policing practices. It also suggests that mapping did certain kinds of cultural and political work in linking the construction of knowledge, its circulation, and the material transformation of places and regions.

The resulting geo-politics of U.S. colonial science (as I have labeled them) hence worked on several levels. On the one hand, an array of scientific bureaus participated in the geographical (and ethno-geographical) politics of territorial division and control, contributing to the construction, definition, and authentication of territorial sovereignty, "habitats," and ethnic identity alongside the development and regulation of new resource extraction regimes. On the other hand, the knowledge produced, communicated, and circulated cartographically served as a resource of (traditional) geopolitical value, and could be mobilized around particular geopolitical positions (notably, retention), depictions of the Philippines and Filipinos for American audiences, and expert and popular debates over U.S. imperialism in the Philippines. And yet of course, the fantasies of imperial control that the U.S. remapping of the Philippines embodied would not be fully realized. Whatever practical and ideological value the maps provided for state projects of territorial transformation, cartographic knowledge alone provided no guarantees against the persistence of difference and "fugitive landscapes" that exceeded state control.[97] But the intersection of Insular science, map-making, and geo-politics surveyed in this chapter does at least allow the *explosion* of mapping in the Philippines after 1898, which Quirino observed,[98] to make sense as traces of a cartographic colonial state attempting to know the territories it presumed to rule.

Notes

1 Padre José Algué, *Atlas de Filipinas/Atlas of the Philippine Islands*, Special Publication No. 3 (Manila and Washington: Observatorio de Manila, 1899; and U.S. Coast and Geodetic Survey, 1900).
2 The volume is available in full on Internet Archive: https://ia800202.us.archive.org/14/items/AtlasPhilippine00Algu/AtlasPhilippine00Algu.pdf.
3 Schurmann to Hay, September 11, 1899. Worcester Philippine Collection Vol. 17, pp. 227–230, University of Michigan Special Collections Library (UM-SCL), Ann Arbor, Michigan.
4 H.S. Pritchett, "Introduction: Origin of the Atlas," in Algué, *Atlas*, p. 3.
5 Schurmann to Hay, September 11, 1899; United States Commission in the Philippine Islands, Expenses in the work: Atlas de Filipinas (contract) received September 1899, Observatorio Central de Manila and the United States Commission in the Philippine Islands, Worcester Philippine Collection Vol. 17, p. 226, UM-SCL. At the same time, the Americans agreed to publish another product of Jesuit geography under Algué's direction, *El Archipélago Filipino: Colleción de Datos Geográficos* (Washington, DC: Imprenta del Gobierno, 1900), a richly illustrated, geographically organized gazetteer published in two volumes. One reviewer, while expressing satisfaction with the Jesuits' work of enumeration, their noteworthy seismic data, as well as the Atlas, judged that the volume as a whole would fail to satisfy either general or scientific readers, with the chapter on ethnology flagged as "distinctly disappointing." Francis H.H. Guillemard, "The Philippine Islands," *The Geographical Journal* 19 (1902): 619–622.
6 Cited in C. Quirino, *Philippine Cartography, 1320-1899* (Amsterdam: N. Israel, 1963).
7 On the intersection of wax engraving, Rand-McNally, and American popular cartography, see Susan Schluten, *The Geographical Imagination in America, 1880-1950* (Chicago: University of Chicago Press, 2002), pp. 17–44.
8 Quirino, *Philippine Cartography*, p. vii.
9 Hence, for Lefebvre, representations of space—the conceptualized spaces (and spatial models) of scientists, urbanists, planners, and technocrats, including maps, plans, blueprints, and landscape designs—were important precisely because they served as prescriptive, not merely mimetic, interventions. Henri Lefebvre, *The Production of Space*, trans. Donald Nicholson-Smith (Oxford: Blackwell, 1991).

10 Benedict Anderson, *Imagined Communities*, 2nd edn. (London: Verso, 1991), pp. 163–186; Winichakul Thongchai, *Siam Mapped: A History of the Geo-Body of a Nation* (Manoa, HI: University of Hawaii Press, 1997); Matthew H. Edney, *Mapping an Empire: The Geographical Construction of British India, 1765-1843* (Chicago: University of Chicago Press, 1997); Matthew G. Hannah, *Governmentality and the Mastery of Territory in Nineteenth-Century America* (Cambridge: Cambridge University Press, 2000); Vicente L. Rafael, *White Love and Other Events in Filipino History* (Durham, NC: Duke University Press, 2000), pp. 19–51; David Harvey, "Cartographic Identities: Geographical Knowledge under Globalization" in Harvey, *Spaces of Capital: Towards a Critical Geography* (Edinburgh: Edinburgh University Press, 2001), pp. 208–234; J.B. Harley, *New Nature of Maps: Essays in the History of Cartography* (Baltimore, MD: Johns Hopkins University Press, 2002); John Pickles, *A History of Spaces: Cartographic Reason, Mapping and the Geo-Coded World* (London: Routledge, 2004); Raymond B. Craib, *Cartographic Mexico: A History of State Fixations and Fugitive Landscapes* (Durham, NC: Duke University Press, 2004); Stuart Elden, "Governmentality, Calculation, Territory," *Environment and Planning D: Society and Space* 25 (2007): 562–580; Scott Kirsch, "The Allison Commission and the National Map: Towards a Republic of Knowledge in Late Nineteenth-Century America" *Journal of Historical Geography* 36 (2010): 29–42; Jeremy Crampton, "Cartographic Calculations of Territory" *Progress in Human Geography* 35 (2011): 92–103.

11 Foucault, from the *Security, Territory, Population* lectures (2004), quoted in Elden, "Governmentality," p. 564. The late colonial state has been defined as an apparatus wherein state institutions became ever denser in their intrusions into social life, seeking to control population, territory, and social interactions in their entirety, yet lacking the resources to do so effectively. J. Darwin, "What Was the Late Colonial State?" *Itinerario* 23 (1999): 73–82; Stephen Legg, *Spaces of Colonialism: Delhi's Urban Governmentalities* (Oxford: Blackwell, 2011).

12 Michel Foucault, "Questions on Geography" (interview with *Hérodote*, January 1976) in C. Gordon (ed.), *Power/Knowledge* (New York: Pantheon, 1980), pp. 63–77, p. 69.

13 On the latter pair, see Warwick Anderson, *Colonial Pathologies: American Tropical Medicine, Race, and Hygiene in the Philippines* (Durham, NC: Duke University Press, 2006); Warwick Anderson, "Science in the Philippines" *Philippine Studies: Historical and Ethnographic Viewpoints* 55 (2007): 287–318.

14 Notwithstanding Lefebvre's inclusion of Foucault among those "speculative philosophers who have diluted the concept [of power] by finding it all over the place...," my premise in the chapter is that the differences between a Lefebvrian approach to maps and colonial state space and a Foucauldian sensibility of distributed power and knowledge are not insurmountable, and indeed may usefully inform one another. While both theorists, in explaining complex spatial arrangements of power, engaged closely with the work of representations in the world, Foucault, Lefebvre insists, had forgotten "where power has its 'real' seat: in the state, in constitutions and institutions." In Neil Brenner and Stuart Elden, "Introduction. State, Space, World: Lefebvre and the Survival of Capitalism" in Lefebvre, *State, Space, World* (Minneapolis, MN: University of Minnesota Press, 2009), p. 12.

15 War Department, *Reports of the Taft Philippine Commission* (Washington, DC: U.S. Government Printing Office, 1901). While the Bureau of Forestry hired a staff 48, including 5 Americans and 43 Filipinos ("Natives, Spaniards, or Chinese"), the Bureau of Mines would employ just four Americans and four Filipinos in its first year.

16 Worcester, then a 32-year-old University of Michigan ornithologist, leveraged two prior scientific visits to the archipelago—and the rapid publication of a 500-page tome, *The Philippine Islands and Their People* (New York: Macmillan, 1898)—to gain appointment to the First Philippine Commission by President McKinley as a regional scientific expert. He remained on the Philippine Commission until 1913, the longest serving (and among Filipinos, most widely reviled) commissioner. For biographical studies, see Rodney J. Sullivan, *Exemplar of Americanism: The Philippine Career of Dean C. Worcester*, Michigan Papers on South and Southeast Asia 36 (Ann Arbor, MI: Center for South and Southeast Asian Studies, University of Michigan, 1991); Mark Rice, *Dean Worcester's Fantasy Islands: Photography, Film, and the Colonial Philippines* (Ann Arbor, Michigan: University of Michigan Press, 2014).
17 Dean C. Worcester, *The Philippines Past and Present*, Vol. I (New York, The Macmillan Company, 1914), p. 491.
18 Paul C. Freer, "Description of New Buildings," Department of the Interior, Bureau of Government Laboratories Circular No. 22 (Manila: Bureau of Printing, 1905), p. 12. American Historical Collection, Rizal Library, Ateneo de Manila University. See also Freer, "The Bureau of Government Laboratories for the Philippine Islands, and Scientific Positions Under It" *Science* 16 (1902): 579–580.
19 Anderson, "Science in the Philippines." See also Jonathan Victor Baldoza, "Under the Aegis of Science: The Philippine Scientific Community before the Second World War" *Philippine Studies: Historical and Ethnographic Viewpoints* 68 (2020): 83–110.
20 Philippine Commission Act No. 156, Section 2, in Anderson, *Colonial Pathologies*, 111.
21 Worcester, *Philippines Past and Present*, Vol. 1, pp. 488–500.
22 Anderson, *Colonial Pathologies*, p. 5.
23 Ibid., p. 113, pp. 113–114. Emphasis added.
24 A.J. Cox, "The Philippine Bureau of Science," Bureau of Science Press Bulletin No. 87 (Manila: Bureau of Printing). Worcester would later claim to have realized what "was commonly referred to as 'Worcester's dream,'" in the Ermita district, describing the construction of Manila's scientific institutions and landscape as "a golden opportunity to start right. In imagination I saw a Bureau of Science for scientific research for routine scientific work, a great General Hospital, and a modern and up-to-date College of Medicine and Surgery, standing by side by side and working in full and harmonious relationship." Worcester, *Philippines Past and Present*, Vol. 1, p. 491, p. 490.
25 Winichakul, *Siam Mapped*, p. 130.
26 Senate Doc. No. 145, 58th Congress, 3rd Session, "Scientific Explorations of the Philippine Islands," pp. 1–22, read and referred to committee in February, 1905, includes the National Academy report (and preliminary research design) and President Roosevelt's message of introduction. RG 350/Box 223/File 1818, National Archives and Records Administration (NARA), College Park, Maryland.
27 Philippine Commission, "Excerpt from minutes of the Commission of September 9. 1903," RG 350/Box 223/File 1818, NARA; Philippine Commission, Excerpt from cable (8465-5) Edwards to Taft, 10/31/1905, RG 350/Box 223/File 1818, NARA. The Commission's endorsement, as noted below, was itself a qualified one.
28 Roosevelt, in Senate Doc. No. 145, p. 1; Taft to Roosevelt 1/20/1905, RG 350/Box 223/File 1818, NARA.
29 Roosevelt, in Senate Doc. No. 145, p. 1.
30 Taft to Roosevelt 1/20/1905.

31 Philippine Commission, "Excerpt from minutes of the Commission of September 9. 1903."
32 Freer to Edwards 2/20/1905, RG 350/Box 223/File 1818, NARA.
33 "The National Academy," Davis wrote of the "objectionable form" of its organizational recommendations for the Philippine surveys, was "NOT the Academy, please remember." Davis to Edwards 1/31/1906, RG 350/Box 223/File 1818, NARA; Davis to Edwards 4/22/05 RG 350/Box 223/File 1818, NARA; Edwards to Davis 4/26/1905, RG 350/Box 223/File 1818, NARA.
34 John Cloud, *Science on the Edge: The Story of the Coast and Geodetic Survey from 1867-1970* (Washington, DC: National Oceanographic Data Center, NOAA Central Library, 2007). Annual reports of the Philippine Department of the Interior and Bureau of Science, and after 1906, the Manila-based *Philippine Journal of Science*, were the primary venues for publication of scientific research on the Philippines among Insular state scientists.
35 George P. Ahern, "Special Report Covering the Period April 1900 to July 30 1901," Forestry Bureau, Philippine Islands, Division of Insular Affairs, War Department (Washington, DC: Government Printing Office, 1901). The U.S. Forest Service was not established until 1905, although a Division of Forestry had been created under the Department of Agriculture in 1881, renamed the Bureau of Forestry under the direction of Gifford Pinchot in 1901.
36 Babette P. Resurrección, "Engineering the Philippine Uplands: Gender, Ethnicity, and Scientific Forestry in the American Colonial Period," *Bulletin of Concerned Asian Scholars* 31 (1999): 13–30.
37 Ahern, "Special Report." The expansion of commercial timber markets in the Philippines under Spanish rule during the second half of the nineteenth century is observed in Greg Bankoff, "One Island Too Many: Reappraising the Extent of Deforestation in the Philippines Prior to 1946," *Journal of Historical Geography* 33 (2007): 314–334.
38 Ahern, "Special Report," p. 6.
39 Ibid., p. 10.
40 Greg Bankoff, "First Impressions: Diarists, Scientists, Imperialists and the Management of the Environment in the American Pacific, 1899-1902," *Journal of Pacific History* 44 (2009): 261–280.
41 Ibid., p. 267.
42 Ibid., p. 268; see also Bankoff, "Deep Forestry: Shapers of the Philippine Forests" *Environmental History* 18 (2013): 523–556.
43 In Bankoff, "First Impressions," p. 270.
44 Alfred W. McCoy, Francisco A. Scarano, and Courtney Johnson, "On the Tropic of Cancer: Transitions and Transformations in the U.S. Imperial State," in McCoy and Scarano (eds.), *Colonial Crucible: Empire and the Making of the Modern American State* (Madison, WI: University of Wisconsin Press, 2009), pp. 3–33; Greg Bankoff, "Breaking New Ground? Gifford Pinchot and the Birth of 'Empire Forestry' in the Philippines, 1900–1905" *Environment and History* 15 (2009): 369–393; Bankoff, "Conservation and Colonialism: Gifford Pinchot and the Birth of Tropical Forestry in the Philippines," in McCoy and Scarano (eds.), *Colonial Crucible*, pp. 479–488; see also J.R. McNeill, "Introduction: Environmental and Economic Management," in McCoy and Scarano (eds.), *Colonial Crucible*, pp. 475–478.
45 George P. Ahern, "Forest Map of the Philippine Islands," Bureau of Forestry, Manila (1910). RG 330/21/18/5-1, Cartographic and Architectural Section, NARA.
46 Bankoff, "Breaking New Ground," p. 386.
47 Resurrección, "Engineering the Philippine Uplands," p. 23.
48 Cloud, *Science on the Edge*. As Cameron Forbes observed shortly after his arrival in the Philippines, the cartography was "mostly done by Filipinos who do most exquisite work. They are charting the whole archipelago, which is a

huge job and has never been done." William Cameron Forbes, "Journal" (entry 9/05/1904), First Series, Vol. I, p. 61, W. Cameron Forbes Papers (1930), MS Am 1365, Houghton Library, Harvard University.

49 United States Coast and Geodetic Survey, "Philippine Islands: Sketch of General Progress" (map) June 30, 1910. Accompanies Report of Supt., 1909-1910. Library of Congress, Geography and Mapping Division, Washington, DC.

50 Edwards to Civil Governor, 8/11/1902, RG 350/Box 272/File 2356, NARA; Cf. United States Coast and Geodetic Survey, "General Progress Sketch: Philippine Islands; Hawaiian Islands; Porto Rico" (map) 1901. Library of Congress, Geography and Mapping Division, Washington, DC; United States Army Signal Corps, "Progress Map" (map) 1903. Library of Congress, Geography and Mapping Division, Washington, DC; United States War Department, "Instructions for a Progressive Military Map of the Philippine Islands" No. 1022855, July 13, 1905, in Ainsworth to Governor-General, July 13, 1905. RG 350/Box 272/File 2356.

51 Hence, the cartographic anxiety identified here differs from the phrase coined by geographer Derek Gregory to evoke the disruption of objectivist distinctions, in geographical representation more broadly, between subject and object, observer and observed; in the context of U.S. Insular government mappings, the action of the former upon the latter, through cartographic renderings, was precisely the point. Gregory, *Geographical Imaginations* (Oxford: Blackwell, 1994).

52 Lanny Thompson, "Governmentality and Cartographies of Colonial Spaces: The 'Progressive Military Map of Puerto Rico,' 1908–1914" in A. Goldstein (ed.), *Formations of United States Colonialism* (Durham, NC: Duke University Press, 2014), pp. 289–315, p. 302. Through progressive military mapping, Thompson (p. 312) argues, "colonial spaces are created through the deployment of techniques of sovereignty, discipline, and governmental rationality. In the U.S. imperial formation, these techniques were widely, yet unevenly dispersed throughout the colonies."

53 Division of Mines, Bureau of Science, "Map Showing Principal Mineral Districts" 1907. Library of Congress, Geography and Mapping Division, Washington, DC. A handwritten annotation "Bureau of Science – Manila – 1908" contradicts the date of 1907 printed on the map.

54 Thompson, "Governmentality and Cartographies of Colonial Spaces," p. 312.

55 Paul A. Kramer, *The Blood of Government: Race, Empire, the United States, & the Philippines* (Chapel Hill, NC: University of North Carolina Press, 2006), p. 5.

56 Rafael, *White Love*, p. 35. The census was under the direction of a U.S. Army General Sanger, who had previously supervised insular census reports for Puerto Rico and Cuba.

57 Ibid., pp. 19–51.

58 Kramer, *Blood of Government*.

59 On the Smithsonian's Bureau of Ethnology (as it was initially called) during the late nineteenth century, see Curtis Hinsley, *The Smithsonian and the American Indian: Making a Moral Anthropology in Victorian America*, 2nd edn. (Washington, DC: Smithsonian Institution Press, 1994); Scott Kirsch, "John Wesley Powell and the Mapping of the Colorado Plateau, 1869-1879: Survey Science, Geographical Solutions and the Economy of Environmental Values" *Annals of the Association of American Geographers* 92 (2002): 548–572.

60 David P. Barrows, "Report of the Chief of the Bureau of Non-Christian Tribes for the Year Ending August 31, 1902," Department of Interior, Manila (1902), pp. 679–688, p. 679. Newberry Library, Chicago.

61 Ibid.

62 Barrows, "Report of the Chief," p. 687.

63 Barrows does not mention an Army or Constabulary escort, though presumably one would have accompanied the Bureau's party at least as far as Kayapa, with the Benguet governor.

64 Ibid.
65 Ibid. The word *Igorot* (or *Igorotte*), from the Tagalog term for mountaineer or mountain people, has persisted as a relatively inoffensive "umbrella" term for the peoples of northern Luzon, though it tends to blur distinctions between varied ethnic and linguistic groups. Existing in complex relation with kin networks, the *rancheria* settlements may better characterize the scale and location of Igorot political communities than the "tribal."
66 P.N. Abinales, "Progressive-Machine Conflict in Early-Twentieth-Century U.S. Politics and Colonial-State Building in the Philippines" in J. Go and A. Foster (eds.), *The American Colonial State in the Philippines: Global Perspectives* (Durham, NC: Duke University Press, 2003), pp. 148–181; P.N. Abinales and Donna J. Amoroso, *State and Society in the Philippines* (Lanham, MD: Rowman & Littlefield Publishers, 2005); Gerard A. Finin, *The Making of the Igorot: Contours of Cordillera Consciousness* (Quezon City, Philippines: Ateneo de Manila University Press, 2005); Renato Rosaldo, *Ilongot Headhunting, 1883-1974: A Study in Society and History* (Palo Alto, CA: Stanford University Press, 1980).
67 Driver, *Geography Militant: Cultures of Exploration and Empire* (Oxford: Blackwell, 2000); see also N. Jardine, J. Secord, and E. Spary (eds.), *Cultures of Natural History* (Cambridge: Cambridge University Press, 1996).
68 Barrows, "The Bureau of Non-Christian Tribes: circular of information," December, 1901, p. 9. Department of Interior, Bureau of Non-Christian Tribes, Manila. Newberry Library, Chicago. The Circular notes that a longer "syllabus" was under preparation by Worcester, based on John Wesley Powell's (1880) *Introduction to the Study of Indian Languages*, for use among ethnological field workers in the Philippines.
69 Barrows, "Bureau of Non-Christian Tribes," pp. 9–10.
70 Ibid., p. 14.
71 Bureau of Science, Division of Ethnology "Directions for Ethnographic Observations and Collections," Manila (1908) Bureau of Printing. National Library of the Philippines, Manila.
72 Rice, *Dean Worcester's Fantasy Islands*, pp. 80–117; Julie A. Tuason, "The Ideology of Empire in National Geographic Magazine's Coverage of the Philippines, 1898–1908" *Geographical Review* 89 (1999): 34–53.
73 See Sullivan, *Exemplar of Americanism*, pp. 141–190; Benito M. Vergara, *Displaying Filipinos: Photography and Colonialism in Early 20th Century Philippines* (Quezon City, Philippines: University of the Philippines Press, 1995); Kramer, *Blood of Government*, pp. 87–158; David Brody, *Visualizing American Empire: Orientalism & Imperialism in the Philippines* (Chicago: University of Chicago Press, 2010); Rice, *Dean Worcester's Fantasy Islands*).
74 James Henderson Blount, *The American Occupation of the Philippines, 1898-1912* (New York: Putnam's Sons, 1912).
75 A. Hoen, "The Philippines: Distribution of Civilized and Wild Peoples," (map). No date (~1900–1902). Library of Congress, Geography and Mapping Division, Washington, DC; Philadelphia Commercial Museum, "Philippine Islands: with Tribal Divisions," (map) 1902. Library of Congress, Geography and Mapping Division, Washington, DC.
76 Bureau of Insular Affairs, "Map of the Philippines," (1904) War Department. RG 330/21/18/5-1, Cartographic and Architectural Section, NARA.
77 Resurrección, "Engineering the Philippine Uplands," p. 17. On the continuing development of U.S. ethnology as a geostrategic project during the Second World War and its aftermath, see Matthew Farish, *The Contours of America's Cold War* (Minneapolis, MN: University of Minnesota Press, 2010), pp. 51–99.
78 Resurrección, "Engineering the Philippine Uplands," p. 19.

79 Timothy Mitchell, *Rule of Experts: Egypt, Techno-Politics, Modernity* (Berkeley, CA: University of California Press, 2002), p. 6.
80 War Department, "Instructions for a progressive military map of the Philippine Islands," No. 1033855, July 13, 1905 in Ainsworth to Governor General, July 13, 1905, RG 350/Box 272/File 2356, NARA.
81 J. Brian Harley, "Rereading the Maps of the Columbian Encounter" *Annals of the Association of American Geographers* 82 (1992), pp. 522–536, p. 532.
82 Some of its officers were known to employ a hybrid of local and prescribed Insular diplomatic practices as a means of resolving clan-based disputes while augmenting their own local authority. Frank L. Jenista, *The White Apos: American Governors on the Cordillera Central* (Quezon City, Philippines: New Day Publishers, 1987).
83 Worcester to Forbes, 10/06/1911. Worcester Philippine Collection, Vol. 21, Part 1, File 9, UM-SCL.
84 Ibid. Such adventurism did not go uncontested within the Insular state. One high ranking Constabulary officer, for example, responded that while "It may be that higher authorities will consider the means were justified by the results, but … I will refuse to furnish escorts to Lieutenant–Governors who take with them hordes of warriors for the purpose of devastating the country." In ibid.
85 Matthew Edney has described practices of mapping and reconnaissance in British colonial India as an *Enlightenment archive* of geographic information under ongoing revision and refinement, in Edney, "Reconsidering Enlightenment Geography and Map-Making: Reconnaissance, Mapping, Archive" in D.N. Livingstone and Charles W.J. Withers, *Geography and Enlightenment* (Chicago: University of Chicago Press, 1999), pp. 165–198.
86 On the historic work of the Constabulary in the Cordillera Central, see Rosaldo, *Ilongot Headhunting*; Finin, *The Making of the Igorot*; Jenista, *The White Apos*.
87 A.S. Guthrie, Bureau of Constabulary General Orders No 21, May 28, 1906. RG 350/Box 272/File 2356, NARA.
88 Ibid.
89 Bruno Latour makes a related point, charting the cartographic emergence of Sakhalin Island in relation to European "centres of calculation," in Latour, *Science in Action: How to Follow Scientists and Engineers through Society* (Cambridge, MA: Harvard University Press, 1987), pp. 215–257.
90 Harbord to Forbes, 3/13/1912 and *Maps of the Philippines by Provinces* (Philippine Constabulary), W. Cameron Forbes Papers, MS Am 1192.8, Houghton Library, Harvard University.
91 Harbord to Forbes, 3/13/1912.
92 *Maps of the Philippines*.
93 Harbord to Forbes, 3/13/1912. The notion of a "geographical gift" is re-purposed here from Michael T. Bravo, "Ethnological Encounters" in N. Jardine, J.A. Secord, and E.C. Spary (eds.), *Cultures of Natural History* (Cambridge: Cambridge University Press, 1996), pp. 338–357.
94 Ibid.
95 Harley, "Rereading the Maps," p. 532.
96 Foucault, "Questions on geography," p. 69.
97 Craib, *Cartographic Mexico*.
98 Quirino, *Philippine Cartography*, vii.

3 Landscape

The Burnham Plans and American Landscape Imperialism in Manila and Baguio

> The power of landscape does not derive from the fact that it offers itself as a spectacle, but rather from the fact that, as mirror and mirage, it presents any susceptible viewer with an image at once true and false of a creative capacity which the subject (the Ego) is able, during a moment of marvelous self-deception, to claim as his own.[1]

Anticipating the December 1904 visit of the prominent Chicago architect and urban planner Daniel Burnham to the Philippines, Cameron Forbes, freshly appointed as Secretary of Commerce and Police, wrote Burnham, an old family friend, a warm personal letter of gratitude and welcome. "I find the Commission greatly pleased at the prospect of your coming," Forbes shared, but he also communicated some of the bleaker realities of colonial life in the Philippines that Burnham could expect to encounter: "Just now things are in a rather depressed condition owing to the recent change in the currency and to the fact that we have just got through an insurrection, a serious pest which killed off a large proportion of the draft animals, and the ravages of cholera and plague." And yet, Forbes continued with a bracing technocratic optimism, "the authorities sanitary, financial and military have so well mastered the problems with which they were confronted that the Islands are more peaceable than ever before in their history, the people have a greater degree of independence of action, and the health here is excellent, the only really troublesome disease being amoebic dysentery, which can be avoided by drinking only bottled waters, and avoiding fresh vegetables, oysters, etc. They have learned also how to cure it," Forbes added, but confided that "the treatment is pretty drastic and very unpleasant."[2]

Burnham's visit had been orchestrated to produce a major plan of proposed improvements for Manila and an original plan for Baguio, the piney mountain village and would-be resort town at the western edge of the Cordillera Central, which the Philippine Commission had already declared the "summer capital of the Archipelago."[3] Whether or not, as Taft insisted, regular visits to Baguio really helped to keep the dysentery at bay, the Burnham plans were expected to address some of the problems of empire by aesthetic means, through interventions in landscape and built

DOI: 10.4324/9780429344350-4

environment. If "things were in a depressed condition," for Forbes and the Insular Government, then the Burnham plans would uplift them, meanwhile leaving an enduring stamp—and perhaps entrenching U.S. geopolitical and economic interests—in the dual Philippine capitals.[4]

Burnham was by this time the celebrated master planner of Chicago's 1893 "White City" World's Columbian Exposition, and well known for the skyscraping and Beaux-Arts achievements of his Chicago architectural firm, Burnham and Root. Contemporaneously with the Philippines projects, Burnham contributed to major City Beautiful planning efforts that included Cleveland, San Francisco, and Washington, DC, later followed by his enduring 1909 plan of Chicago. In these designs, Burnham had taken lessons from the blend of monumental neoclassicism, the emerging field of landscape architecture, and circumscribed public spaces that had been "crowd-tested" in the temporary, festival spaces of the White City, for developing more or less permanent projects of urban landscape transformation.[5] Given this track record in the production of spectacular urban spaces, Burnham's involvement in the American effort to remodel Manila was a matter of prestige for the Insular Government, and in particular for Forbes, the Boston Brahmin grandson of Ralph Waldo Emerson—and heir to the accrued fortunes of the Forbes China trade and railroad capital—who had drawn on his own networks of cultural capital to enlist the architect.[6] The opportunity to mark the emergence of American empire in tropical Asia was evidently an attractive one for Burnham, who like Forbes easily adopted the Republican Party aura of the reluctant imperialist. Or not so reluctant. When Forbes, prior to his own departure for Manila in July 1904, had asked the architect's advice on who to hire for the Manila and Baguio projects, Burnham, perhaps divining Forbes's intentions, recommended himself.

Burnham would steam into Manila Bay with a designer from his firm, the architect Pierce Anderson, on 7 December 1904. The two remained in the Philippines for about six weeks, carrying out site visits around Manila and initiating work on the plans; meanwhile Burnham was feted by Insular officials and military leaders. Burnham and Anderson also journeyed with Forbes—by train, carriage, and horseback—to Baguio, located 233 kilometers north of Manila and roughly 1,540 meters *up* from sea level, where a 10 square-mile reservation had been set aside (by the Commission) within which the summer capital plan was to take shape. The Plan of Proposed Improvements for Manila, submitted to Congress from Chicago in June 1905, would situate Manila, the Spanish colonial capital since 1571, within an evolving American planning tradition at a moment when North American urban spaces were themselves being intensively reconstructed.[7] In the Philippines, the Burnham plans would also serve to place landscape aesthetics squarely on the agenda of U.S. cultural imperialism and geopolitics. The Manila Plan would present, alongside a bayfront landscape enhanced for elite consumption, at least a quasi-democratic distribution of City Beautiful public spaces and greenways. But the continuing investment in Baguio—which required construction and maintenance of a road ascending 5,000 feet by steep switchbacks to a largely American enclave of mountain cottages, playing fields, sanitarium, and military camp, and soon Forbes's own magnificent Topside residence, a modern stone bungalow perched majestically on a ridge overlooking Baguio and surrounding mountains and valleys, complete with stables, gardens, and guest quarters—appeared as singularly tone-deaf to local conditions. On the heels of the decade of devastating warfare, famine, and disease to which Forbes referred in his 29 August letter, the high costs of building the summer capital at Baguio (and maintaining access to it despite persistent monsoon washouts and road closures) would open the Commission to criticism on both sides of the Pacific.

Taking Burnham's 1904–1905 visit to the Philippines as a starting point, this chapter examines efforts to extend American empire *through* landscape, focusing on aesthetic dimensions or what might be called a landscape vision of U.S. empire. Its purpose is to understand how the ideological contradictions of the imperial moment—between democracy and empire, liberator and subjugator—were built into American colonial spaces, sometimes brutally but sometimes through aesthetic means in the formation of setting and landscape. As proposed interventions in landscape, the Burnham plans offer glimpses of the linked spatial and symbolic strategies for structuring social encounters and everyday relations of power in the Philippines—or in the case of Baguio, for attempting to geographically rise above them. Before rejoining Burnham and Anderson on Manila Bay in 1904 to more closely examine their work in the Philippines, the next section turns briefly to the historical, conceptual convergence of *landscape* and *aesthetics* which was a key premise of their work so as better situate the aesthetic landscape within broader efforts to reproduce American empire in the Philippines.

The Aesthetic Landscape

Landscape is commonly taken for granted to mean the (usually scenic) setting for human experience, or the representation of such settings in painting and other visual arts. In traditional geographical research, landscape emerged as a key morphological concept describing lands as shaped by both nature and human practices into differentiated regional settings, each with a distinctive regional economy and "look of the land."[8] This highly visualized convergence of aesthetics and landscape, which may appear as perfectly "natural," is suggestive as to how the landscape idea is practically intertwined with the history of aesthetics—by the time Burnham arrived in Manila to gaze upon the walled *Intramuros* for the first time and lay out his vision for the American colonial capital, the construction of park spaces and built environments (increasingly known as "landscapes") *without* explicit consideration of aesthetics would have been unthinkable.

If, as Kant interpreted the field from classical Greek roots, aesthetics constituted a specialized engagement with the material world as apprehended by the senses, a science of sense perception, then by the mid-nineteenth century, the term was gaining currency in English in more explicit connection with visual appearance and its human effects, especially in relation to beauty and the arts.[9] Landscape, as a site for sensory apprehension and the experience of beauty, grandeur, and symbolic elements, thus emerged as a conceptual category alongside the professionalization of landscape aesthetics and expertise. Perhaps initially through landscape painting, Europeans "represented their world as a source of aesthetic enjoyment—*as* landscape."[10] Europeans were hardly the only ones to extend what might be called an aesthetic landscape ideal, evoking beauty, order, and harmony, into the material environment of the landscape "itself" in landed estates and gardens.[11] But it was an explicitly aesthetic sensibility of landscape, embraced in Europe in new urban architectures and public parks, that Burnham cultivated in his own planning efforts, in connection with the new discipline of landscape or landscape architecture in late-nineteenth-century North America (most prominently reflected in Frederic Law Olmstead's work) and its incorporation into City Beautiful urban planning.[12]

For geographer Denis Cosgrove, taking landscape as a historically constructed *way of seeing* helped to open the aesthetic landscape to a potent critique of ideology. Thus, the development of landscape painting, for Cosgrove, thinking of the sixteenth-century Palladian villas of the Veneto, was "intended to serve the purpose of reflecting back to the powerful viewer, at ease in his villa, the image of a controlled and well-ordered, productive and relaxed world wherein serious matters are laid aside."[13] Whether painted on canvas or sculpted into the material environment, aesthetic landscapes have often served to erase the conditions of their own production, or to naturalize a particular "order of things," especially at moments of political (territorial) or property transition.[14] The Burnham plans for Manila and Baguio would offer cover, respectively, for both, setting the tone for an American landscape imperialism that would become more widely distributed across the archipelago in the following decade.[15] To recognize the political dimensions of landscape aesthetics is not, of course, to reject the value of aesthetics in built environments, to reduce the stakes of urban planning to their aesthetic dimensions, nor even to foreclose the possibilities of social uplift through beautification that animated City Beautiful planners. But such perspectives offer tools for raising critical questions about the prioritization of aesthetics by an influential regime of Insular government actors who, as we see in this chapter, were deeply invested in what might be called a landscape vision of U.S. empire in the Philippines.

In telling this story, the chapter engages with "social formation and symbolic landscape," after Cosgrove, in the context of early U.S. colonial state interventions in the Philippines, and attempts to read these landscapes "through" Lefebvrian categories of spatial production and state theory.[16]

Lefebvre understood space as a multifaceted product of contested social relations; similarly he saw the state itself as existing in persistent tension with social forces that threatened to undermine it at "weak points," withering away an always unstable authority. This authority was especially precarious in colonial and imperial contexts in which the state lacked legitimacy. Reading the aesthetic landscape through Lefebvrian categories allows us to examine the Burnham plans not only as "representations of space"—the plans themselves—or in terms of their concrete outcomes (and contemporary traces) in the physical landscape, but also as moments in a process of seeing, interpreting, and reconstructing spaces intended to reflect the interests and values of those who produced them. It compels us, in this context, to be attuned to registers of beauty and delight that were central to their production and function *as* landscapes. And yet while landscapes may be designed as naturalizing or aggrandizing symbolic spaces, their meanings, let alone their capacities for channeling social behavior, are not inherently stable, even as landscape iconography is often fashioned precisely to evoke a sense of permanence or timelessness.[17] Hence, the stabilization of meaning in form is precisely the cultural work—and aesthetic politics—that produced landscapes are intended to achieve.

In the Philippines, efforts to create a distinctively American colonial landscape at the start of the twentieth century, while also creating a set of landscapes distinctly *for* Americans, were prioritized by a small "aesthetic regime" of elite Insular state actors as pivotal problems of Philippine governance, an essential cultural politics of landscape and built environment. The fledgling summer capital at Baguio, located near the site of an earlier Spanish Army garrison and sanitarium at La Trinidad, most closely embodied this aesthetic. Also inspired by the British "hill station" at Shimla, India, advocates deemed a summer health resort at Baguio to be vital for Americans living in the tropics as a space not only for surviving the hardships of colonial life in the tropics, but also for enjoying its beauty and pleasures, signaling the aesthetic registers on which the American empire was to be experienced by its agents abroad.[18] To understand how the Burnham plans were produced and, in different ways, realized as American colonial spaces, in the next section I turn more closely to the social relations through which the political and aesthetic project of American landscape imperialism was forged, including intimate, embodied relations of cultural authority, race, nation, class, and gender, and particular kinds of relationships, like friendship, in which meaning, information, and "common sense" were easily shared.

Insular Architecture

Plans for the creation of Baguio as an American hill station and summer capital—and for improving access to the fledgling mountain resort—were well underway in February 1904 when the 34-year-old Forbes, after meeting once with President Roosevelt and twice with Taft in Washington, accepted the position of Secretary of Commerce and Police in the Insular Government with a seat on the Philippine Commission. In addition to "legislating for the Islands" as a Commissioner in "all except a few matters in which Congress has retained the power, as foreign relations, currency, tariff, etc.," Forbes's portfolio was nearly as broad as the colonial state itself. It would include supervision over the police, coast guard, coast and geodetic survey, road and harbor improvement, corporations and franchises (including railroads and light and power companies), the post office, the government-owned telegraph, and, as Forbes recorded in his journal, "the building of the new summer capital at Benguet."[19] Forbes quickly became devoted to the latter task, helping to secure US$ 375,000 for the Benguet Road in the commission's preliminary annual budget in September 1904, and visiting Baguio and the Benguet Road worksites and labor camps beginning the month following his arrival in the Philippines.[20] Forbes would continue to prioritize Baguio in different ways—and to channel extensive resources toward the road's maintenance—throughout his Philippine career.

While Taft, in the initial interview, told Forbes that he wanted a businessman with railroad experience for the position, it is clear that Forbes's social cachet—a "great clubman as well as athlete" from an elite Boston family (see Figure 3.1)—was also attractive for a government still struggling to generate new U.S. and international investment in the Philippines.[21] Roosevelt urged him "to take this place; it is doing some of the world's work," Forbes noted, while Taft advised him to think in terms of a ten-year commitment in the Philippines.[22] Forbes, who had earlier unsuccessfully sought appointment to the U.S. Panama Canal Commission, would accept, succeeding Luke Wright (who was promoted to Governor-General following Taft's call to Washington) as Secretary of Commerce and Police. Interestingly, while Baguio was mentioned among Forbes's new responsibilities after the meeting with Taft, the Manila plan was not, suggesting that Baguio may have been the "tail that wagged the dog" in the American planning efforts.

Among his administrative duties, Forbes would also take it upon himself to guide the development of a coherent American architectural style during his tenure in the Philippines; Forbes sought to turn the U.S. regime into an explicitly aesthetic one. As Forbes later reflected on the state of government architecture at the time of his arrival in Manila:

> The office of the Insular Architect was <u>a scandal</u>. The man in charge was without taste or tact, his work <u>hideous</u>, and the result was that all of the government buildings and structures were done without the artistic eye and architectural sense that should be evidenced by a civilized people as an example in dealing with less civilized peoples. I couldn't arouse very much enthusiasm on the part of the Commission in this attitude and the transfer of the office to my department was necessary in order to achieve any architectural fitness in our work, because with

ARRIVAL OF W. CAMERON FORBES, COMMISSIONER

Says When Bill Now Before Senate Is Passed Plenty Capital Will be Found For Railroads Here.

GREAT CLUBMAN AS WELL AS ATHLETE

Figure 3.1 The arrival in Manila of new Philippine Commissioner W. Cameron Forbes was hailed in the pages of the *Manila Cable News* (9 August 1904).

> an unsympathetic secretary the architect would have a pretty hard time when he ran up against the hard headed bureau chiefs, each of whom wanted his ideas carried out in regard to his structures—school buildings, customs houses, provincial buildings, hospitals, etc., etc. Unless the secretary of the department supported the architect and insisted upon architectural beauty and good design of buildings, the utilitarian bureau chief would have been prevailing and no good architect would have been willing to remain in a position where his recommendations were being constantly ignored or turned down.[23]

Hence, for Forbes, not only did the American colonial landscape lack coherence it was also a missed opportunity in the imperial endeavor, an unrealized moral and political force for dealing with "less civilized peoples." Burnham's visit to the Philippines was an early milestone in these efforts, establishing Manila and Baguio as symbolic sites where these goals of architectural demonstration might be achieved while also improving the quality of life for an American colonial cadre secure in the knowledge that it was "doing the world's work." Burnham had earlier recommended Forbes for the Panama Canal Commission, but the opportunity to work in the Philippines, Burnham had insisted, would allow Forbes to be "constructive in a higher sense."[24] After Forbes, in turn, helped to bring Burnham to the Philippines, the two would work closely together in efforts to realize this sensibility in new urban plans for Manila and Baguio.

Manila's urban makeover, of course, was by no means only a matter of aesthetics. As architect Gerard Lico has argued, under the Municipal Board during the early years of U.S. rule, "a succession of epidemic diseases attributed to the unhygienic practices of the Filipinos" was used to justify an "aggressive overhaul of the physical urban space and the native practices that occurred within its domain."[25] American public health interventions, broadly celebrated in the pages of *National Geographic*, were experienced in the Philippines with brutal specificity at times through new forms of state violence and bureaucracy, organized around an emerging biopolitics of sanitation and hygiene that included extensive burning of nipa palm houses in infected areas, confiscation and destruction of foodstuffs, and forced bathing of inhabitants in bichloride solution.[26] Lico describes "highly meticulous and militaristic sanitary squads" in Manila that "moved through the city, street-by-street, quarter-by-quarter, house-by-house, accounting the daily census of the living and dead, arresting suspected and infected native bodies for medical incarceration, the cleansing and burning of contaminated nipa houses, and the removal and disposal of the dead by cremation boundaries."[27] While many new municipal codes, such as Ordinance No. 21, "prohibiting the practice of cleaning ears, scraping eyelids or barbering on the streets, lanes, alleys and public squares," attempted to control what were seen as unruly behaviors, others targeted the built environment itself, attempting to regulate the construction and repair, and mandated whitewashing, of buildings, and the use of materials in traditional (nipa) houses.[28] If urban public space in particular constituted a "highly ordered colonial space – a controlled landscape shielded from potentially disruptive social forces," then, the symbolic and material elements were deeply entwined in such projects.[29] Landscape's aesthetic dimensions—wherein the designers of "landscape," may seek, in geographer Don Mitchell's words, to "substitute the visual for the (often uncomfortable and troublesome) heterogeneous interactions of urban life"—were internal to its material transformations.[30]

While the Commission's authority in the Philippines depended on the War Department, its capacity to transform the Philippines along the lines it envisaged was also undermined from Washington, since the U.S. Congress, animated by diverse strains of American nativism, protectionism, and anti-imperialism, would not authorize the removal of key tariffs regulating U.S.–Philippine trade until 1909, after a decade of U.S. colonial governance. Perhaps precisely because of these political and economic constraints, the cultural politics of landscape and built environment took on particular significance for the evolving Taft-Forbes regime. With U.S. troop levels having been reduced, in 1903, below 70,000 (including the southern archipelago), the Burnham plans also signaled a deepening U.S. commitment to its presence in the Philippines and Asia, one to be realized "concretely" in the space of planned urban landscapes and built environments. Against a range of challenges, the regime would continue to invest itself in the terrain of aesthetics, setting the tone for U.S. colonial policy in the Philippines under successive Republican Party administrations until 1913.

During Taft's governorship, the Americans had moved quickly to begin building the resort in Baguio, starting with a military hospital and sanitarium along with a handful of cottages and private homes for top government officials. Before Filipinos were even represented on the Commission, it authorized a continuing stream of funds for the completion and maintenance of the Benguet Road (later Kennon Road), despite a bewildering array of setbacks and cost overruns, offering one clear measure of the priority afforded to the summer capital idea.[31] For key actors in the making of Insular state policy, Baguio came to be seen as central to the U.S. imperial endeavor in the Philippines, a distinctively ordered aesthetic and recreational space rising above the heat, humidity, and contentious politics of Manila. Boosters of the "Philippine Adirondacks" retreat justified expenditures on Baguio (and the Benguet Road) in terms of a therapeutic and morally invigorating landscape, a space to which Americans could "return," making possible more extended and productive stays in the tropics.[32] Enacting this vision of colonial life pivoted not only on large-scale geographical (or geopolitical) forces of history but also on the work of a relatively small network of actors who were effective, for a time, in constructing and holding together a formal aesthetic vision of American colonial life in the Philippines, to be achieved in part through conventions of landscape.

Making Plans

> Found Mr. Burnham at my house and was delighted to see him. First time I've been called Cam since I left San Francisco.... Talked all day with Mr. Burnham, telling him conditions and plans and progress and driving him about the city.[33]

Insisting that there was "no decent hotel in Manila," Forbes had invited Burnham and Anderson to be his houseguests during their visit. Burnham would have horses and carriage at his disposal, Forbes offered, along with "Chinese servants, who understand looking after the personal comforts of guests."[34] Linking this distinctive racial economy of labor with corporeal comforts, Forbes's sense of intimacy and shared social context are evident in the exchange. Enjoying tropical sunsets, late suppers, perhaps brandy and cigars, Forbes's guests would not lack the opportunity to exchange views with their host about conditions in the Philippines. Forbes's reflection on being "called Cam," upon returning "from a trip up the islands" to find Burnham installed at his home in Manila points us to specific locations wherein meaning, values, and perspectives on the Philippine landscape could become shared. The feelings of comfort and familiarity; the capacity to spend all day (among many others during the visit) talking amiably, telling of conditions, plans, and progress, and driving around the city together; and the social and cultural connections that made their friendship possible, enhanced their collaborative capacities through the development of shared perspectives on the Philippine landscape and polity.

Even before his journey across the Pacific, however, the excitement over the summer capital idea had rubbed off on Burnham, for it was Baguio, not Manila, on which he appeared most eager to work. As Burnham described it for the Chicago *Daily Tribune* in an article celebrating American efforts to remake Manila as the "Gem of the Far East" and Baguio as the "Philippine Simla":

> The Manila problem is one of suiting new plans as far as possible to existing conditions. I am, therefore, more deeply interested in the summer capital project. There I shall be able to formulate my plans untrammeled by any but natural conditions. I shall make an extended study of the buildings best suited to the climate and other demands of Philippine life, and then go ahead with the creation of a brand new city.[35]

Travelling to Luzon to embark on this work, Burnham combined the personal and professional, boarding the ocean liner *Mongolia* in San Francisco on 13 October 1904 with his wife Margaret Sebring Burnham, 20-year-old daughter Margaret, and several Chicago friends. After six days, the Burnhams disembarked briefly at Honolulu, followed by 11 days at sea to Yokohama. There the party boarded a train for Tokyo the next day, initiating a four-week holiday in Japan, touring its famed temples and gardens in Tokyo, Nikko, and Kyoto. Back in Yokohama, Daniel Burnham, along with Pierce Anderson, would board the British liner *Doric* en route to Manila, while Mrs. Burnham and Margaret set course back to Honolulu to await Daniel's return.

Steaming into Manila Bay on 7 December, the architect jotted a quick note to his wife, noting that he had been met by a government launch and honored with "courtesy-of-the-port" (no customs inspection). "The place is set against the sea," Burnham observed broadly, with lenses trained on the imperial moment, "We can look out over the grounds where Dewey fought at Cavete [sic]."[36] With Forbes as his host, Burnham would find himself enchanted with Manila, albeit by qualities that rarely could be attributed to Filipinos themselves, such as its tropical setting or the "distinctly Spanish" architecture of Intramuros.[37] Burnham was also charmed with the vision of American colonial life presented to him in his travels. Together with Anderson, a Beaux-Arts trained designer, and Forbes, who effectively represented the Insular state as the "client" in the relationship, Burnham initiated work on both the Manila and Baguio plans during his visit, designing spaces for the reproduction of that vision through interventions in landscape that were both practical and symbolic.

Burnham's diary suggests an active and, for a thickset 58-year-old gentleman unaccustomed to the tropical climate, perhaps exhausting schedule of activities. Drives around the city with Forbes were just the beginning. On 9 December, Burnham enjoyed a minstrel show and midnight supper on board the battleship *Wisconsin* with Rear Admiral Yates Sterling, commander-in-chief of the U.S. Asiatic Fleet, then was up early the next morning "seeing members of the Navy and working over the problems of the new city plan."[38] On 11 December, he rose at dawn to travel with Anderson and Forbes up the Pasig River by navy tug and across the Laguna de Bay, transferring to dug-out canoes before heading upriver to a waterfall, "each man paddled by two Filipinos," later arriving for a "sumptuous dinner" with Insular officials in Pasig Town at 10 PM.[39] On 14 December, after a visit to Bilibid Prison, Burnham and Anderson spent the "afternoon over the plan of Manila" in an office set up for them in Manila's Municipal Building, then back for dinner on the *Wisconsin*; on 15 December Burnham inspected an "insane hospital" with Interior Secretary Dean Worcester. From this point, however, Burnham appears to struggle periodically with illness and fatigue during the visit. We know that a doctor was called in to see him at Forbes's home on 18 December, prescribing something, but that Burnham was apparently well enough to be a guest of honor for dinner at the home of Governor Wright on 19 December and to travel on 20 December, rising at 4:30 AM to board a special train north en route to Baguio with Forbes and Worcester.

The journey to Baguio was no doubt a demanding one for Burnham. It required two full days (and one night) of travel, including an arduous final climb by horseback up the rough and unfinished Benguet Road. After lunch at the railroad station at Dagupan—the northern extent of the Spanish line—the travelers, who included Worcester's wife Nanon and daughter Alice, Forbes's steward, Yu Dong, a cook, and perhaps additional service staff, and at least one Constabulary officer, travelled by two four-horse teams that carried them as far as Camp Four off the Benguet Road by evening. At Camp Four the party enjoyed an evening of dinner and native performances at the workers' camp hosted by Major Lyman W. V. Kennon, the army engineer called in to overcome problems of labor, geology, weather, and graft that had plagued the road's early construction efforts. The next day entailed the more difficult journey up the steep switchbacks that the Benguet Road would become famous for in postcard views. Forbes and Worcester left at sunrise to get a head start on arrangements in Baguio while Burnham and others in the party slept in, taking on the ride after lunch. They arrived at the small sanitarium in Baguio in time for dinner at seven.

Figure 3.2 Daniel Burnham takes in the montane landscape during his December 1904–January 1905 visit to Baguio.

Source: Moore (1921), *Daniel H. Burnham*.

The architects would spend nine days in Baguio spanning the Christmas holiday, settled at Governor Wright's summer house (see Figure 3.2)—one of just a handful of homes to have been constructed for American visitors in the presumptive summer capital—and taking meals at the sanitarium. But while Forbes would look back on "a fine week in Baguio" full of the "vim and vigor Baguio gives the red corpuscles," Burnham on 22 December 1904 would confess to feeling "slightly uneasy inside," his activities correspondingly reduced.[40] We know from Forbes that Burnham enjoyed a "wonderful walk" over the town site on that day ("filled with promise for fine development") but that it was just Anderson with whom Forbes shared a "wonderful ride" a few days later when the pair "selected a half dozen visionary house sites," perhaps including what would become Forbes's own signature Topside residence, with its spectacular Cordillera views.[41]

Working closely with the architects, Forbes was excited by the project of creating Baguio in the mind's eye before building it in the landscape:

> It is new, strange, and delightful, this laying out a city on this rolling land. We have about thirty things to provide sites for and we want to distribute them beautifully, conveniently and expediently ... hotels, official buildings, schools, sanitarium, playgrounds ... country club, etc. The problems are great fun; water supply, how to make the little really level ground go round, etc.[42]

In Baguio, the priority given to the problem of aesthetics (for Forbes, the delicious, Burnhamesque challenge of distributing things beautifully, conveniently, and expediently) was not only a local issue but also a broader "Insular" one, leveraged through the Philippine Commission. The City Beautiful emphasis on iconic landscape views, surfaces, and façades and the creation of Baguio itself as a recreational space for the American colonial cadre were to be prioritized over different problems—and different road-building needs—that might have been addressed by the Insular Government. Built, as Forbes conveyed, from the desire for beauty, health, and pleasure among "familiar" alpine surroundings, Baguio and the Benguet Road emerged together as hallmarks of an explicitly aesthetic regime.[43]

On 30 December, the party remounted for the trek down the unfinished zigzag of the Benguet Road, arriving back at Kennon's Camp Four for dinner but leaving Burnham, miserably, "too tired to sleep."[44] After returning to Manila by train from Dagupan the next day, Burnham would spend the remainder of his visit in more genteel spaces—dinners along the riverfront with the Governor-General, and on board naval flagships—along with a visit to Cavite, while continuing work with Anderson on the Manila plan. By early January 1905, Burnham had already produced, Forbes believed, "what seems to me a stunning plan of Manila. It will take many dozens of years and millions of dollars, and won't be completed while any of us are alive, but we can make a start on it."[45] Before leaving Manila, Burnham and Anderson also furnished sketches that would "permit immediate work" on several projects, including a plan of the proposed seafront boulevard and Luneta extension ("New Luneta"); a sketch layout of proposed streets on the harbor front to facilitate a new military reservation; and a sketch street layout of Malate, "cutting" the Malate military reservation.[46] The Baguio plan surfaces in Burnham's journals after he had departed Manila when Burnham notes that he and Anderson had been "working on the Baguio scheme" while on board ship from Hong Kong to Yokohama 21–24 January. Returning to Osaka and Kyoto, Burnham reported that the pair had completed a preliminary sketch map of the City of Baguio (which Burnham sent to Forbes from Kyoto with an explanatory note). He went on to meet Mrs. Burnham and Margaret in Honolulu as planned on 13 February, finally putting back to Chicago via San Francisco on 9 March. "The Manila scheme is very good," he told one confidante, but it was the Baguio scheme that began "to warrant hope of something unusual among cities."[47]

Accompanied by the highly detailed Plan of Manila and a smaller-scale plan of Manila Bay featuring a proposed sea boulevard, Burnham and Anderson's *Report on Proposed Improvements at Manila* was submitted in its final form to Secretary of War Taft (and to Forbes in Manila) from Chicago on 28 June 1905.[48] The previous day the preliminary plan of Baguio had been completed; the full report and town plan for the summer capital—which required additional on-site survey data to be completed—would be submitted in October.[49] As we examine in the next section, these plans (or, in Lefebvrian terms, representations of space) provided Forbes and the Insular Government with organic frameworks for the (re)fashioning of Baguio and Manila. In turn, their selective realization in the landscape "itself" reflected distinctive sets of U.S. colonial priorities as well as "weak points" in the Insular state's capacity to transform Philippine spaces.

Representations of Space: Manila

> Improvements of great scope are obtainable in Manila by reasonable means. On the point of rapid growth, yet still small in area, possessing the bay of Naples, the winding river of Paris, and the canals of Venice, Manila has before it an opportunity unique in history of modern times, the opportunity to create a unified city equal to the greatest of the Western world, with [the] unparalleled and priceless addition of a tropical setting. In keeping pace with the national development and in working persistently and consciously toward an organic plan in which the visible orderly grouping of its parts one to another will secure their mutual support and enchantment, Manila may rightly hope to become the adequate expression of the destiny of the Filipino people as well as an enduring witness to the efficient services of America in the Philippine Islands.[50]

It is not difficult to parse Burnham's soaring rhetoric: while Manila might hope to *become* an adequate expression of the destiny of the Filipino people, in the meantime his plans for improvements, developed with Anderson, emphasized the city's functions as a colonial capital and imperialist landscape, and thus, unavoidably, also reflected the contradictions of an empire of democracy or "empire of no empire." The plan for Manila featured a revamped street system with radiating arteries extending diagonally across the city, monumental government buildings, and public parks, and a new bayfront esplanade of public–private spaces, designed as the new center of public life in the capital (see Figure 3.3). Moving these core functions outside

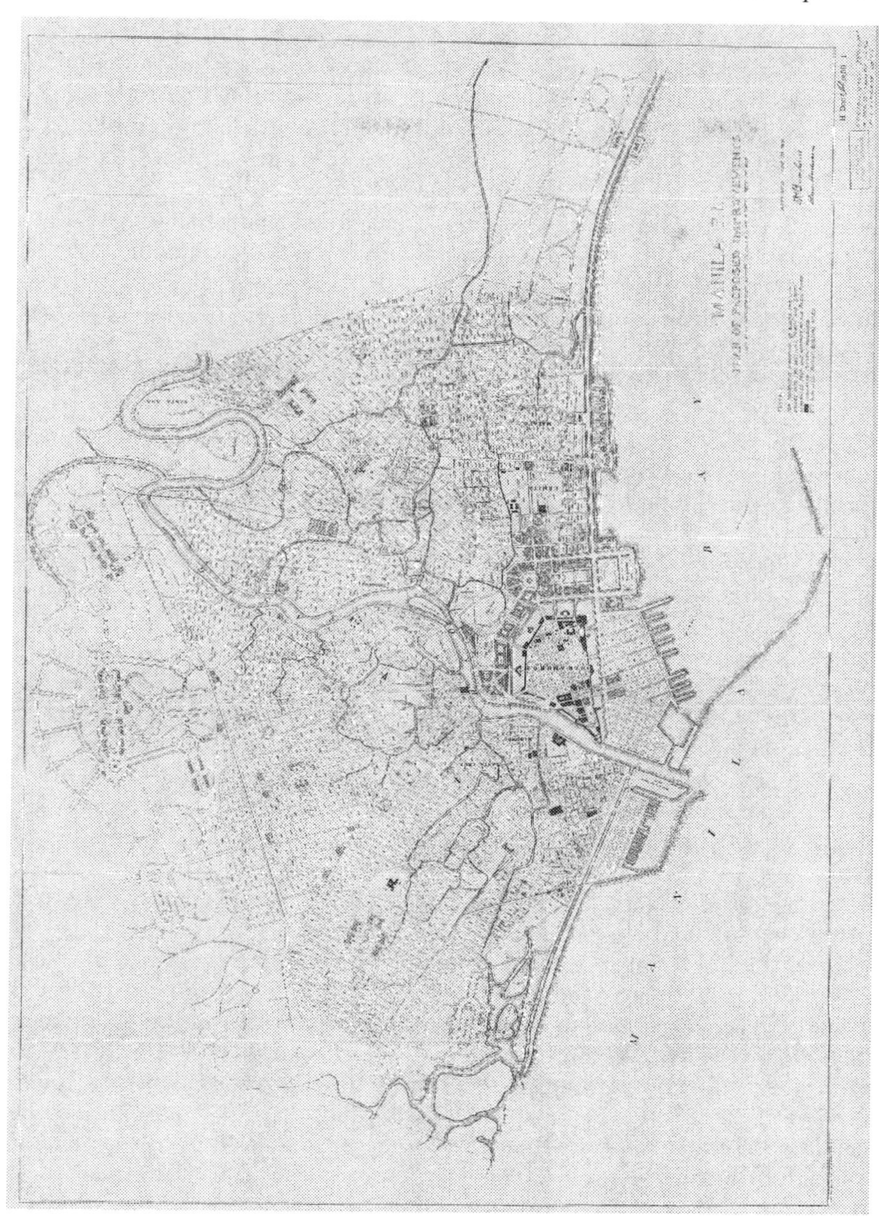

Figure 3.3 Daniel Burnham and Pierce Anderson's (1905) "Manila P.I. Plan of Proposed Improvements," 28 June 1905, with the walled Intramuros visible in the lower center of the plan alongside the port expansion, New Luneta, Government Group of Buildings, and bayfront development including hotel, casino, and private clubs.

Source: Courtesy, Geography and Map Division, Library of Congress.

Intramuros, as planning historian Ian Morley argues, "had great symbolism as it accentuated the fact that the Intramuros, the spatial embodiment of Spanish colonization and military might … was no longer the cultural and governmental nucleus of the country."[51] The Burnham plan offered a new city of open views, set apart from the walled space of the fortress city, by forming monumental vistas along the city's new boulevards, government buildings and monuments in landscape settings, and spectacular views onto Manila Bay.[52] As an organic framework, however, the development of the plan, as Forbes recognized, would be a deliberate, incremental process, integrating with elements of the extant landscape.

While the Manila plan shared characteristics with the temporary exposition grounds of Chicago and St. Louis, site of the 1904 Louisiana Purchase Exposition, Burnham's report also reflected an appreciation of aspects of Manila's historic landscape and was attuned to particular, fine-grained local details. He expressed delight with Intramuros buildings of "a distinctly Spanish type … built of wood with projecting second stories; and their screen windows were built of translucent shells set in a small mesh grille. The roof, which still further overhangs the building, was commonly covered with beautiful dull red tile, and the effect of the whole is unusually pleasing."[53] The "general effect of the existing well shaded narrow streets is picturesque," Burnham affirmed, "and should be maintained."[54] His proposed street system for Manila would leave the old city streets relatively untouched, except for the creation of the "wide diagonal arteries."[55] Predicated on a new round of practical and symbolic investment in the built environment, the Plan of Improvements was meant to be implemented gradually, expanding on existing urban features to make a relatively minimal intervention that could be stamped, nonetheless, as modern, Western, and American. For Burnham as for Forbes, the American redesign of Manila anticipated a landscape around which both imperial and democratic futures could be imagined.

Described in terms of Manila's "landscape possibilities," Burnham's use of buildings, landscape architecture, and natural qualities to create spaces as pictorial settings—as landscape—surfaces repeatedly in the Manila plan in the form of plazas, circles, esplanades, and parkway boulevards, and in smaller informal park spaces.[56] Outside the old walls of Intramuros, a sunken garden would be created by backfilling the moat, leveling off the defensive counterscarp so that a "proper setting" would be formed for the old city while also removing what was understood, by Manila's new American occupants, as a malodorous hazard.[57] Meanwhile, opening pedestrian access to the tops of the old walls reflected another distinctly visual approach, turning residents and visitors to the city into viewers from above, adding a "unique touch to the monumental aspect of the town."[58] As for the natural setting, while the intense heat of the tropics posed physical challenges, the local environment also provided an elemental means of refreshment through visions of water in the physical landscape:

> Besides the possibility of abundant foliage and fountains of water, Manila possesses the greatest resources for recreation and refreshment in its river and its ocean bay. Whatever portions of either have been given up to private use should be reclaimed where possible and such portions as are still under public control should be developed and forever maintained for the use and enjoyment of the people.[59]

The sensibility of the city *as* an aesthetic landscape became predominant in the design for the New Luneta (Luneta Extension) and other public and semipublic spaces along Manila Bay, even as what counted as public control, or even as enjoyment of the people, had been interpreted quite narrowly in the plan, in the interest of the colonial class.

While part of the Spanish-built Luneta, including José Rizal's execution site, had been pledged for a memorial space, much of the "present" Luneta, as Burnham called it, was to be occupied by a Government Group of buildings, cultural institutions, and park spaces.[60] Together with the reworked public spaces and new architectures of democracy, the production of *new* lands in Manila Bay, including the "New Luneta," would constitute the symbolic and functional core of Burnham's City Beautiful approach to improving Manila. The New Luneta was to be built 305 meters farther out in the bay through land reclamation, using backfill dredged from the harbor as part of a major port expansion. The New Luneta would spoil the old Luneta's view of Manila Bay, but in this movement of public space into the bay the plan would restore to the Luneta its "former commanding outlook, partially cut off by the new works of the port, and also to form a larger pleasure park near the center of town and on the water front."[61] Opening onto the bay, the monumental landscape of the Government Group, although pointedly understated compared with much neoclassical architecture of the era, was an object lesson intended to produce demonstrable moral effects. As Burnham described the designated Hall of Justice, "Its architectural expression should speak the greatness of its function. The moral effect of such a hall of justice, magnificent in outward form and aspect, compelling an attitude of respect, if not inspiring a feeling of awe, would be cheaply secured at large sacrifices of space and money."[62]

While the Manila plan signaled inspiration from several European sources along with the local environment, it was modeled concretely after two North American cities in the midst of their own City Beautiful transformations: Washington and Chicago. Washington, for Burnham, was "the best planned of all modern cities—by superposing a system of wide diagonal arteries on the rectangular system above described. These arteries, with the radial ones springing from the center, all of them being wider than the average street permit parklike connections with space for trolley cars, reaching all important centers of the city."[63] Emphasizing circulation and flow, the diagonal arteries in the Manila plan would similarly "reach out to all sections of the city."[64] The plan included a complete numbering of city blocks, creating a new "street system securing direct and easy communication from every part of the city to every other part," though Burnham understood that such a street system would take many years to be realized.[65] This presumptive democratization of the built environment, it was argued, should extend "to give proper means of recreation to every quarter of the city."[66] While "the use of parks as an architectural accessory has long been common," Burnham argued, "it has remained for the modern city, with its immense and congested population, to show the necessity of them as breathing places for the people."[67] Hence, just as Chicago's South Park Board had recently approved 14 new public parks and playing fields, including four parks up to 32 hectares in size (each containing a club building, hall for public entertainment, reading rooms, gymnasia, baths for men and women, and small swimming pools), the Manila plan allocated "nine such parks, evenly distributed over the city."[68] In addition, summer resorts and country clubs would ring the hill country around Manila's city limits.

The plan's signature "breathing place" was along the waterfront, where elite clubs and residences (including the Governor-General's) and a world-class luxury hotel were to be built on new land between the bay and esplanade, forming the "natural theater of the social life of Manila."[69] Featuring open parks and boulevards intended to contrast sharply with the closer spaces of the Spanish capital within Intramuros,[70] the production of the new bayfront as a landscape of leisure and prosperity was one that the architects felt compelled to justify ideologically even as they laid out the terms of their design:

> Stretching south from the governor-general's residence, also on new-made land, extend a series of city clubs, whose character as semipublic institutions justifies giving up to them a portion of the water front. Each club will have ample grounds for gardens and outdoor games, as well as a broad terrace on the seaward side with suitable planting for protection from the sun's glare and the typhoon. It is believed that the close grouping of these clubs, as in London, will enhance their value to the whole community. The concentration of social activities through the related grouping of official residences, hotels, and clubs in parkway boulevards and gardens along the water front will, it is believed, make possible an attractive social life that will bring many influential people to Manila and count for much in the prosperity of the islands.[71]

Thus, an elite social landscape could be rendered a matter of colonial responsibility for the Insular state. Forbes had already emphasized for Burnham the demand for a luxury, Western-style hotel, and what would become the Manila Hotel would indeed be prioritized in planning the city's New Luneta and bayfront esplanade. Henceforth, a site was laid out in the Burnham plan on new land adjacent to the New Luneta park, "reserved for a hotel whose size, surroundings, and appointments are intended to deliver

Manila once and for all from the standing reproach of inhospitality toward a traveler."[72] While the plan's monumental spaces had been envisioned as moral agents for shaping Philippine subjects, the bayfront was to be reproduced as a space for the aesthetic pleasures of viewing the bay, for seeing others and being seen, and for contemplating the geographical setting. "The delightfulness of a city," for Burnham, was thus "an element of first importance to its prosperity, for those who make fortunes will stay and others come if the attractions are strong enough, and the money thus kept at home added to that freely spent by visitors will be enough to insure continuous good times. The aim should be to make Manila, really, 'The Pearl of the Orient.'"[73] The waterfront landscape envisioned in the Manila plan was a space designed to reflect on its producers—the political class—precisely its own sense of self-worth.

Representations of Space: Baguio

For Baguio, Burnham and Anderson delivered City Beautiful landscape urbanism at a smaller scale, laying out the town site (for a population not expected to exceed 25,000) over the Insular Government's recently set-aside reservation. The Baguio plan featured an elegantly distributed gridiron of streets, government buildings, landscaped park settings, and other green spaces laid out across the valley's mountain ridges, steep slopes, and hilly terrain, with municipal and national government groups of buildings perched on opposing plateaus above the town's commercial center (see Figure 3.4). Although somewhat more ornate than the Manila plan, here too Burnham's aesthetic reflected values of order, efficiency, monumentality, and hierarchy, attempting to naturalize these values in the landscape of the summer capital. As Burnham described it in the June 1905 draft report, the "Baguio plain furnishes the one practical site for business activities, and, while closely connected with the municipal center, will remain subservient to it. The National Group, while in view of business and within easy reach, will nevertheless frankly dominate everything in sight of it."[74] The "other enclosing hills," featuring expansive valley views, would "furnish locations for various semi-public functions whose buildings, of monumental character, will be in view of one another." Further along the ridge, breathtaking mountain vistas to the east and, on a clear day, glimpses of the South China Sea 25 miles to the west, situated the new town in spectacular fashion on the western edge of the Cordillera Central. Outlying slopes, extending away from the valley toward La Trinidad, would be available for private villas and semipublic institutions, including schools, universities, hospitals, and sanitaria, which called for detachment from the town center. "The total effect of this whole arrangement," Burnham offered immodestly, "—the business center, surrounded by a crown of monumental buildings, the whole dominated by the group of National Buildings, could be made equal to anything that has ever been done."[75]

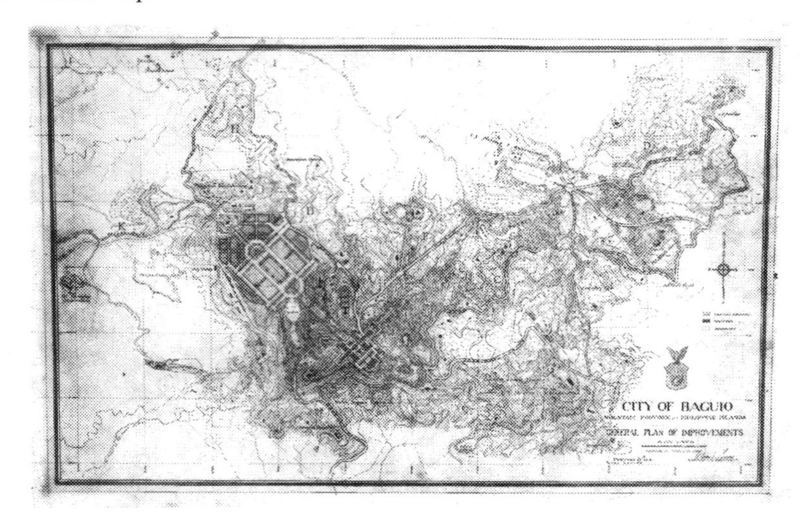

Figure 3.4 Daniel Burnham and Pierce Anderson's Plan of Baguio as effectively realized in William Parsons's (1913) "City of Baguio: General Plan of Improvements," depicting "proposed," "executed" (fully shaded), and "temporary buildings."

Source: Courtesy, Ryerson and Burnham Archives, Art Institute of Chicago.

Burnham invited the plan's readers to stroll through an imagined Baguio, visualizing it already *as* a landscape, a well-ordered view from above, harmonizing efficiency of function with the formation of pleasing civic and recreational settings:

> If one will imagine the long main axis, expanded by an open green esplanade, stretching down from the Municipal group through the business town and up to the green slopes to the dominant government center on the high hills: imagine certain transverse axes, crossing the town and leading up the inclines to important buildings on the flanking hills: look for the green play fields here and there and picture the entire composition hemmed in by the pine ridges of the highest hills, and one will have before him an architectural group of unsurpassed effect and a business machine of the utmost efficiency.[76]

The plan for the Insular summer capital reflected even more starkly than the Manila plan the contradictions of American imperial democracy. For while Forbes would continue to rationalize that Baguio and the Benguet Road should be built "with a view to giving the people of the Philippines access to a temperate climate in a short time," he also intimated that the mountain resort was, for Americans, a *necessity* for thriving in the Philippines.[77] Certainly, the urgency of the Baguio plan and Benguet Road efforts, built to entrench

American interests in the summer capital before plans could be reversed (and to be enjoyed while present colonial careers in the Philippines were ongoing), made plain for whom the landscapes of Baguio were produced *for*, at least in the first instance. As Forbes would put it in a 1909 letter to Secretary of the Navy George von Meyer—in attempting to convince von Meyer to establish a new Naval hospital in Baguio, nearby the polo field, golf links, and country club—"I believe the availability to playgrounds a desirable thing; it makes officers more contented and is a good thing for convalescents when they reach a certain degree of strength ... I am absolutely convinced of the usefulness and desirability of Baguio as a health resort. In my personal case it has made the difference between my being contented out here and otherwise. I do not think I could have done the work and stood the racket that I have if I had not been able to go up into the hills where I could exercise vigorously in the daytime."[78] And yet while the polo field, as Forbes would emphasize for von Meyer, was built at his own personal expense, it occupied a rare swath of flat land in the valley and depended on the infrastructural investments of the Insular Government.

As early as 29 December 1904—Burnham and Anderson were still in Baguio with Forbes, Worcester, and party—the English-language *Manila Times* reported on a "Fence Thrown Around Baguio" in the form of a ten-square mile reservation for the new town site, containing lots to be put up for public sale.[79] The Insular Government's production of space and landscape in Baguio, of course, was never merely a matter of aesthetics and design, pleasure and recreation, but also one of primitive accumulation and land commodification, underwritten by state expenditures.[80] Forbes reflected as much in his journal the following week, excited that "if Mr. Burnham can prepare preliminary plans we can slap in about twenty miles of roads which will make the real estate salable."[81] In May 1905, turning out a land market bolstered by Burnham's prestige (and the promise of completing the Benguet Road), Forbes was delighted to report to Burnham that the first lot sale had been a "tremendous success," with 89 lots in the Baguio reservation sold for residential purposes to "Americans, Englishmen, Spaniards, and Filipinos."[82] These lots included the spectacular hillside setting on which, in his mind's eye—with the aid of a sketch by Pierce Anderson[83]—Forbes had already begun to build his own elegant mountain home and named it Topside.

Visions Realized? The Burnham Plans and the Power of Landscape

Despite Taft's tacit approval, Forbes struggled initially, in the wake of Burnham's visit, to put his new aesthetics in place, taking it personally that the Commission had "turned me down on my plan for fixing up Baguio."[84] But support for the Benguet road persisted. Even after a rainfall of more than 18 inches in 24 hours caused the road to slide out in several places, Forbes managed, at a 1 May 1905 meeting held *in* Baguio, to secure from the Commission "enough money to complete *and maintain* the Benguet road."[85] Developments in Manila were also promising. By the end of June 1905, the Commission had passed the Luneta Extension Bill funding the scheme to use earth dredged from the harbor expansion to produce new land along Manila Bay. "This is practically adopting Mr. Burnham's great plan," Forbes enthused in his journal, "It provides a hotel site, several club sites, and gives the present Luneta and Camp Wallace field for the site of the Insular Government buildings, when we get round to building them. Highly pleased."[86] Burnham and Anderson, with Forbes, understood that key elements of the Manila plan, such as the establishment of new street systems in the old quarter, could be developed only gradually, as the state acquired the necessary private property.[87] Still, that Forbes could take the Luneta Extension alone as constituting the practical adoption of the Burnham plan offers a telling reflection of colonial priorities. The Luneta Extension Bill also indicated Forbes's consolidation of political currency within the Commission, and his willingness to invest that currency in the colonial landscape.

In September 1905, Forbes advanced a resolution through the commission "agreeing to the transfer of the Bureau of Architecture from the Department of Public Instruction to mine" and arranging "for the resignation of the Insular Architect, who is no good, and got authorization from Commission for appointment of [William] Parsons of New York whom Mr. Burnham has arranged for.... This gives us a good architect."[88] By the following May, the new aesthetic regime was secured when—after a cloudburst had spoiled a Country Club luncheon that Forbes was privately hosting for government clerks—the Commission, "in session" at Baguio, passed a bill formally adopting the Burnham plans for Manila and Baguio. The bill established Parsons as "consulting architect for the Commission" and stipulated that "no changes can be made by the city or insular government in the city walls, parks, or buildings without consulting the architect first and getting his opinion in writing." "Parsons is delighted with this," Forbes exulted, "it should put an end to a lot of vandalism that the tasteless American official loves to indulge in. I feel this bill to be a real accomplishment."[89] The development of a coherent American colonial aesthetic, and extension of empire through interventions in landscape, appeared within reach.

Parsons arrived in Manila on 17 November 1905—less than a year after Burnham's visit. By December he had already ventured to Baguio, setting to work laying out a street system in light of the survey made for Burnham's final town plan. In addition to implementing elements of the Burnham plans and serving as the landscape's architectural supervisor, Parsons's eight-year Philippine career would include signature buildings like the Manila Hotel, Philippine General Hospital, and Army-Navy Club in Manila, designs for two- and eight-room schoolhouses and municipal buildings distributed across the archipelago, and new city plans, along Burnhamesque City Beautiful lines, for provincial capitals in Zamboanga and Cebu.[90] The Office of the Consulting Architect, consisting of Parsons, one American and five Filipino draftsmen, was by no means limited in its work to Manila and Baguio, as Parsons's list of ten preliminary and 15 completed building plans—produced between April and June 1906 alone—makes clear (see Table 3.1).[91] Parsons observed, in June 1906, that "few buildings of a permanent character have been erected," and, through the use of reinforced concrete and local hard woods (rather than imported Oregon pine), he sought to address the perceived deficiency—the absence of an enduring American architectural landscape—through durable construction materials as well as elegant design.[92]

Table 3.1 Building Plans by the Office of the Consulting Architect, 5 April to 30 June 1906

Completed Plans and Specifications	
Project	*Location*
Trade School	Zamboanga
Industrial School	Vigan
Manila Inter-Island Transportation Building	Manila
Office for Surveyor of the Port	Iloilo
Trade School	Bacalor
Provincial School	San Juan de Guimba
Constabulary Quarters	Zamboanga
Engineering Building	Baguio
Provincial Building	Tuguegarao
Cottage	Singalong Experiment Station (Manila)
School	Arayat
Provincial School Shops	Tuguegarao
Provincial Building	Albay
Constabulary Officers' Quarters	Malolos
Barrio School	Santa Ana, Pampanga
Preliminary Plans	
Project	*Location*
Manila General Hospital	Manila
Proposed Warehouses	Paseo de Magallanes (Manila)
Provincial Building	Tárlac
Provincial Building	Santa Cruz
Provincial Building	San Fernando
Provincial Building	Aliminos, Pangasinan
Provincial Building	Pasig, Rizal
Market	Zamboanga
Market	Pasig, Rizal
Audiencia (high court) Building Addition	Manila
Constabulary Headquarters and Barracks	Albay
Jail	Malolos

Source: William E. Parsons, "Annual Report of the Consulting Architect for period extending from November 17th, 1905 to June 30th, 1906," 30 June 1906, p. 3. Daniel H. Burnham Collection, Box 56, FF 22. Ryerson and Burnham Archives, Art Institute of Chicago.

Meanwhile, initiated as early as 1903, and expanded after the practical adoption of the Burnham Plan in 1905, millions of cubic yards of earth, dredged from the sea floor to create space for deep-water berthing piers for ocean going steamers (and an expanded Port of Manila), were deposited at the site of Burnham's New Luneta for the production of 148 acres of new made land.[93] Allowing the new land time to settle before major construction could begin, the corner stone for a new hotel—located on Manila Bay on the new Cavite (now Roxas) Boulevard—would be laid by U.S. Secretary of War Jacob Dickinson on a 1910 state visit (see Figure 3.5). Slated to be "the finest hotel in the orient," without equal in the matter of interior decorations, the Manila Hotel, with its exotic flooring of *pagatpat* and *mangachapuy* woods, was both earthquake- and fire-proof while affording, "all the comforts of a private home."[94] Under Parsons's direction, the "permanent character" of the New Luneta, bayfront esplanade, and Government Group took shape during the first years of the 1910s.

Developments in Baguio were in some ways realized more quickly. Forbes had been "particularly anxious" about the Baguio plan, "lest the chance be lost to make the city beautiful."[95] By early May 1906, he believed he had achieved a powerful consensus with Major General Leonard Wood, the military governor of Moro Province, to "get Baguio properly under way with Secretary Taft, himself, and me, three Baguio enthusiasts, representing the three interests. He will rush the military end of it."[96] Within three years of Burnham's Cordillera visit, the Insular Government had moved some operations to Baguio for up to four months per year, with the Commission and its clerks at the vanguard, as colonial bureau chiefs built mountain homes or rented government cottages. Indeed, the letterbooks of Forbes and Worcester reveal that, for leading Insular officials, residence in Baguio was by no means limited to summer seasons.[97] The Baguio Country Club, chartered,

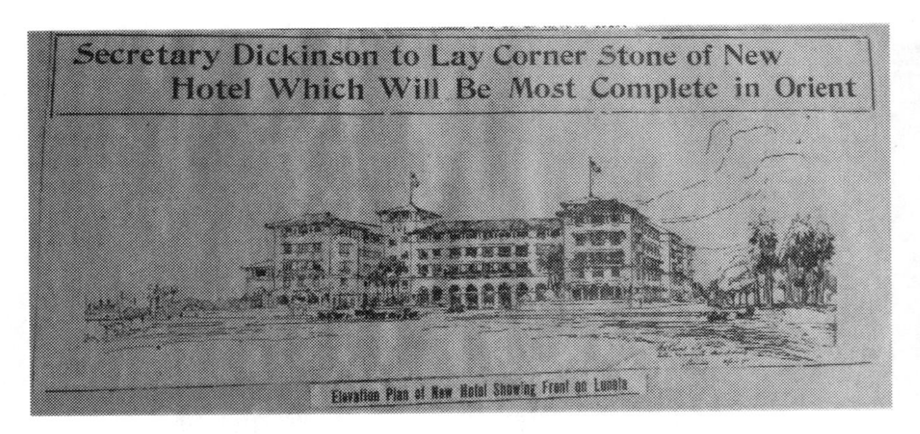
Secretary Dickinson to Lay Corner Stone of New Hotel Which Will Be Most Complete in Orient

Elevation Plan of New Hotel Showing Front on Luneta

Figure 3.5 The Manila Hotel, designed by Insular Architect William Parsons, as appearing on the front page of the *Manila Times* 29 August 1910.

improbably, in 1900 (by Worcester's recollection), and incorporated in 1907, was expanded, with the help of a personal investment from Forbes, from a "rude, grass-roofed shed made of pine slabs" to include a clubhouse, kitchen, and living room, along with cottages and sleeping rooms that could be rented by members.[98] Tennis courts and trap-shooting facilities were installed, and the club's three-hole golf course, famed for its Igorot caddies dressed only in western shirts and g-strings, was expanded to nine links. Yet whatever its rustic origins, and however exculpatory was the patina of "private" investment, the country club was the product of the immense subsidy—based on revenues of the "self-supporting" Insular state—that made Baguio accessible, habitable, and 'ownable.'[99] This logic was not lost on critics of Baguio, like the anonymous cartoonist in a 1911 *Philippine Free Press* editorial depicting Baguio as "our white elephant" (see Figure 3.6), for whom it was only the exploitation of Filipinos and Philippine resources that it was possible, for American colonials, to enjoy Baguio as a genteel, naturalizing landscape of comfortable retreat.[100] Criticism over Baguio, however, whether from the Manila press or anti-imperialists in the U.S. Congress, was a price that its boosters were willing to pay; it was not, in the short term, a politically costly one, judging from Taft's election to the U.S. presidency and Forbes's subsequent promotion to Governor-General.

10 CENTAVOS

Philippines Free Press

10 CENTAVOS

VOL. V. | Philippines Free Press Manila, P. I., Saturday, August 19, 1911.

Our White Elephant | Nuestro "Elefante Blanco"

Figure 3.6 "Our White Elephant" in *Philippine Free Press*, 19 August 1911, p. 1.

Baguio's development under Parsons's authority featured a range of projects: a Government Center would include offices and dormitories for government workers; Camp John Hay, the military sanitarium, was to be enlarged; new homes, along with Methodist and Episcopal churches, were built on lots which Forbes had helped to lay out during his Christmas 1904 visit with Burnham and Anderson. Other facilities were upgraded. As early as May 1905, Forbes had reported to Burnham, in a letter reflecting the inconveniences Burnham had evidently suffered during his visit, that:

> The Sanitarium has … improved four hundred percent. We got a hustling manager, put in a good cook, a lot of Chinese waiters, and that with the opening of the road which brought frozen beef and other delicacies within reach has made the service and fare quite respectable. Then we have opened the new part of the Sanitarium which is much better and the new dining room with netting to keep flies out from the meals, so that many of the aggravations which were present when you were there have disappeared.[101]

Parsons returned to Baguio in April and May 1906, establishing a general system of roads around Baguio—laid out as trails and narrow wagon trails but with 15 meters in width set aside around each as "part of the park system," with the expectation of subsequent road widening.[102] The roads would connect four new "sub-divided" sections around Baguio, comprising 144 new lots sold at public auction 28 May 1906, and others set aside religious, social, and 'semi-public' purposes.[103] As Parsons would report the next month, the auction now made possible "the development of Baguio by private resources."[104] Built largely from wood, not reinforced concrete, Parsons was aware that, for his clients in the Insular Government, Baguio must be produced in time for the present generation of U.S. colonials to enjoy. Parsons also understood precisely that Baguio was to be built *as* a landscape, wherein "the cutting and planting of trees … should be controlled by the principles of landscape art rather than those of forestry. Natural beauty of landscape may be enhanced by the formation of vistas, and contrasts between thickly wooded hillsides and open glades." In this sense, as Parsons understood it, "the development of Baguio and vicinity has become a problem of landscape gardening and architecture rather than one of engineering."[105]

Meanwhile, Baguio's boosters sought to legitimize the venture by extending its benefits to additional groups of government workers. In 1908, "Teachers Camp" was established; there American educators posted to public schools throughout the archipelago could summer together, playing baseball, attending "tent lectures," riding horses or loafing in the high-country meadows of Benguet. For Worcester, the renewal of American identity at Teachers Camp was as essential as the sharing of best pedagogical practices:

> Americans who spend too many years in out-of-the-way municipalities of the Philippines without coming in contact with their kind are apt to lose their sense of perspective, and there is danger that they will grow careless, or even slovenly, in their habits. It is of the utmost benefit for school teachers to get together once a year, learn of each other's failures and successes, and profit by each other's experiences, forget their troubles while engaging in healthful athletic sports, listen to inspiring and instructional discourses, and above all else benefit by open-air life in a temperate region.[106]

The purported benefits of the summer capital were not limited to Americans. After 1908, when the Bureau of Lands began transferring both American and Filipino employees to Baguio for the summer season, Worcester would insist that the "small additional expense involved was more than justified by the larger quantity and higher quality of the work performed as a result of the greatly improved physical condition of the workers."[107] Unlike his speculations about the cultural renewal experienced by American teachers, however, Worcester's observations of Baguio's effects on Filipinos were limited to measures of body mass: "Every Filipino sent to Baguio gained in weight," he boasted, "with the single exception of a messenger who had to run his legs off!"[108] More bureaus were encouraged to follow suit,[109] necessitating new construction, using primarily wood and other relatively impermanent materials, of a new dormitory and mess hall for government workers adjacent to the Government Center, a geometrical landscape perched above the town center in which values of beauty, order, and efficiency were easily recognizable.

Baguio also fired the geopolitical imaginations of American colonials. Forbes and Worcester were among those who advocated for an enduring American protectorate, centering on Baguio and Mountain Province, even in the event of Philippine independence. The region was prized by Worcester, in Department of Interior reports, for its defensible, mountain fortress-like qualities as a space for retreat in the event of insurgency.[110] After the boundaries of Mountain Province were redrawn in 1908 to "include corridors to the sea in the north and the west," even Forbes's successor as Governor-General, Francis Burton Harrison, later wondered whether "this maneuver was designed to 'secure a final separation of these portions of Luzon' from an independent, if not rebellious, Philippines."[111] As Governor-General, Forbes went so far as to propose, in a 1910 letter to Secretary of War Dickinson, a plan, perhaps cooked up with Worcester, "for the creation of a militia up there [based in Baguio], which I believe would go a long way toward assisting us in case the defense of the Islands was necessary, either against invasion or insurrection, besides accomplishing a great good in the sections affected."[112] Already a means of naturalizing American power in the Philippines, Baguio was also a landscape around which different geopolitical futures could be imagined.

Two years later, Forbes would elaborate his vision of an imperial Fortress Baguio in a confidential letter to General John J. Pershing:

> My idea of a mountain militia has been to establish somewhere in the Mountain Province a school at which military training should be taught to a number of selected young men from the tribes which show greatest aptitude for the purpose and which they are to be found in the Mountain Province.[113]

The letter, delivered to Pershing across town while both were resident in Baguio, also incorporated suggestions received from General Bell, suggesting that Forbes's imperial mountain militia had become at least a matter of speculation among leading American officials. Forbes worried, however, about "the possible deleterious effect of the propinquity to civilization of savages who are liable to acquire vices," arguing instead for the establishment of the training school *outside* of Baguio around an arsenal to be situated in the center of Mountain Province. Building loyalty, based on ten-year commitments from the enlisted militiamen, was central to the scheme by which Baguio was to be protected, perhaps hived off, from a space of Filipino sovereignty. Hence, in what would remain, for Forbes, an unrealized imperial fantasy:

> The advantages of the militia would be a people who could not be tampered with; who are loyal to Americans; who have never advocated the withdrawal of the Americans from the Islands, and who could be depended upon to be loyal in case of invasion or insurrection: a people who demand nothing, as they have no particular income, in expecting large pay; who are accustomed to get their own food; who need no transport service; whose uniform would consist of only one upper garment and a gee-string; who are accustomed to carry great weights long distances; who need no camp outfits; who cook for themselves; who can travel over thirty miles of mountain trail in a day carrying with them everything they are likely to use; who are excellent shots, as evidence by their work in the Constabulary.[114]

Meanwhile, the pleasures of imagination extended, for Forbes, to the construction of his own magnificent 'Topside' residence above Baguio. Designed by Parsons based on the original hand-drawn sketches by Pierce Anderson,[115] Topside was a large stone bungalow situated on a ridge above Baguio (see Figure 3.7), offering views, from its ample windows, sun porch, and manicured grounds, of the steep mountain valleys and long ridges of the Cordillera Central in the distance. By 1906, Topside could accommodate 12 guests but Forbes soon added a second story above the kitchen wing with an additional three bedrooms and two bathrooms.[116] It was, presumably, the pinnacle of social invitations, and by April 1907 Forbes was satisfied that

Figure 3.7 In a staged photograph, U.S. Secretary of War William Howard Taft (far right, in hat) confers with his host, W. Cameron Forbes, during a 1907 visit to Topside, Baguio, as their party looks on. From Forbes, *Topside Guestbook* (1906–1912).

Source: Courtesy, Houghton Library (MS 1365.4), Harvard University.

his "Topside book" already included the names of 70 visitors.[117] The guests were predominantly Americans but also elite Filipinos, including Sergio Osmeña, then governor of Cebu, Attorney General Gregorio Araneta, and Philippine Commissioner Benito Legarda, who together "came to live with" Forbes at Topside in March 1907 and were, Forbes surmised, delighted, "all converts to Baguio, if opposed before."[118] Whether entertaining visiting dignitaries or hosting Igorot *canaos*—or both—Forbes and his staff presided over a social scene that was deeply invested in its own aesthetic enjoyment.[119] With several guests typically visiting at any time during summer and holiday seasons, Topside was Forbes's personal manifestation of the power of landscape, a space designed for the apprehension of beauty and, through the intimacy of this experience, for enhancing social and political relationships. Constructed in blue granite quarried from the hill above and offering spectacular Cordillera views from the comfort of the garden, here was a landscape that reflected back on its viewers (from a mountaintop!) their own sense of self-worth. In this naturalized empire of delight, the problem of *distributing things beautifully in space* could be prioritized as a functional element of the American colonial presence in the Philippines, appealing, for

Forbes and other powerful agents of American empire, to a deeply personal sense of national and imperialist identity.

***

This chapter has explored efforts to extend and hold together a formal vision of American colonial life in the Philippines through conventions of landscape and the dream of empire by aesthetic means. Through the lens of an "intimate history" of the production of colonial spaces in Manila and Baguio, I have argued that aesthetic dimensions of landscape were prominent and iconic in these efforts, offering indications of linked spatial and symbolic strategies of U.S. imperialism that were prioritized by a Taft–Forbes regime determined to leave its mark on the archipelago. Reproducing Manila and Baguio *as* landscape, that is, as stable, well-ordered, and harmonious "views from above," would help to make possible the survival and reproduction of an American Philippine empire over time, or so advocates believed. The deception is a tempting one, for the *power* of landscape, as Lefebvre argues in the chapter's epigraph, derives precisely from a "moment of marvelous self-deception" wherein the viewer is able to claim the image as their own.[120]

Notes

1 Henri Lefebvre, *The Production of Space*, trans. Donald Nicholson-Smith (Oxford: Blackwell, 1991), p. 189.
2 Forbes to Burnham, August 29, 1904. Daniel H. Burnham Collection, Box 1, FF 31, Ryerson and Burnham Archives, Art Institute of Chicago, Chicago, IL.
3 US Philippine Commission, "Excerpts from minutes of the Philippine Commission of 1 June 1903," Executive Bureau, Legislative Division, 24 June 1903, RG 350/Box 534. National Archives and Records Administration (NARA), College Park, Maryland.
4 Forbes to Burnham, 29 August 1904.
5 Thomas S. Hines, *Burnham of Chicago: Architect and Planner*, 2nd edn. (Chicago: University of Chicago Press, 1979); Carl Smith, *The Plan of Chicago: Daniel Burnham and the Remaking of the American City* (Chicago: University of Chicago Press, 2006); Christine Ellem, "No Little Plans: Canberra, via Chicago, Washington DC, the Philippines, and Onwards," *Thesis Eleven* 123 (2014): 106–122; Christopher Vernon, "Daniel Hudson Burnham and the American City Imperial," *Thesis Eleven* 123 (2014): 80–105.
6 Hines, *Burnham of Chicago*, pp. 197–216.
7 Daniel H. Burnham and Pierce Anderson, "Report on Proposed Improvements at Manila" Sixth Annual Report of the Philippine Commission, Part 1, pp. 627–635. Bureau of Insular Affairs, War Department. (Washington, DC: U.S. Government Printing Office, 1906).
8 Paul Vidal de la Blache, *Tableau de la géographie de la France* [Portrait of the geography of France] (Paris: Hachette, 1908); Carl O. Sauer, "The Morphology of Landscape," *University of California Publications in Geography* 2 (1925): 19–53.
9 Raymond Williams, *Keywords: A Vocabulary of Culture and Society*. Rev. ed. New York: Oxford University Press, 1983), p. 31.

10 Denis E. Cosgrove, *Social Formation and Symbolic Landscape*, 2nd edn. (Madison, WI: University of Wisconsin Press, 1998), p. 1, emphasis added. See also Cosgrove, "Prospect, Perspective, and the Evolution of the Landscape Idea," *Transactions, Institute of British Geographers* 10 (1985): 45–62; Veronica Della Dora, "Topia: Landscape Before Linear Perspective," *Annals of the Association of American Geographers* 103 (2013): 688–709.

11 Denis E. Cosgrove and Stephen Daniels (eds.), *The Iconography of Landscape: Essays on the Symbolic Representation, Design and Use of Past Environments* (Cambridge: Cambridge University Press, 1988); Stephen Daniels, *Humphry Repton: Landscape Gardening and the Geography of Georgian England* (New Haven, CT: Yale University Press, 1999).

12 Thomas S. Hines, "The Imperial Façade: Daniel H. Burnham and American Architectural Planning in the Philippines," *Pacific Historical Review* 41(1972): 33–53.

13 Cosgrove, *Social Formation*, p. 24.

14 Ibid.; see also Don Mitchell, *The Lie of the Land: Migrant Workers and the California Landscape* (Minneapolis, MN: University of Minnesota Press, 1996).

15 Ian Morley, "Modern Urban Designing in the Philippines, 1898–1916," *Philippine Studies: Historical and Ethnographic Viewpoints* 64 (2016): 3–42; Ian Morley, *Cities and Nationhood: American Imperialism and Urban Design in the Philippines, 1898–1916* (Honolulu, HI: University of Hawai'i Press, 2018).

16 Cosgrove, *Social Formation*. As noted in the Introduction, while the 'production of space' was theorized, in Lefebvre's signature work, as a means of contributing to the survival of capitalism over time, he subsequently turned more explicitly to the pivotal role of the state in these processes. Lefebvre, *Production of Space*; Lefebvre, *State, Space, World: Selected Essays*, eds. N. Brenner and S. Elden (Minneapolis, MN: University of Minnesota Press, 2009). Architect Gerard Lico explored Lefebvrian dimensions of American colonialism in Manila in Lico, "Building the Imperial Imagination: The Politics of American Colonial Urbanism and Architecture in Manila," *Philippine Humanities Review* 9 (2007): 239–270.

17 Jean Gottman, *La politique des états et leur géographie* [The politics of states and their geography] (Paris: Armand Colin, 1952).

18 Dean C. Worcester, *The Philippines Past and Present*, Vol. 1. Reprint, Whitefish, MT: Kessinger Publishing, pp. 253–276; Robert Reed, *City of Pines: The Origins of Baguio as a Colonial Hill Station and Regional Capital* (Baguio City: A-Seven Publishing, 1999); David Brody, *Visualizing American Empire: Orientalism and Imperialism in the Philippines* (Chicago: University of Chicago Press, 2010), pp. 140–163; Rebecca Tinio McKenna, *American Imperial Pastoral: The Architecture of US Colonialism in the Philippines* (Chicago: University of Chicago Press, 2017). For the persistence of 'environmentalist' logic among U.S. geographers after the Second World War, see J.E. Spencer and W.L. Thomas, "The Hill Stations and Summer Resorts of the Orient," *Geographical Review* 38 (1948): 637–651.

19 William Cameron Forbes, "Journal" (February 1904), First Series, Vol. I, p. 3. W. Cameron Forbes Papers (1930), MS Am 1365, Houghton Library, Harvard University. The ten-volume *Journals of William Cameron Forbes* were initiated shortly prior to Forbes's tenure in the Philippines to be shared with his mother Edith Emerson Forbes in Massachusetts. They were also written for posterity. Forbes later lightly annotated the journals, donating the first five volumes along with his papers, for posthumous release, to Harvard libraries in 1930.

20 Forbes, "Journal" (entry 9/04/1904), Vol. I. The visits, and Forbes's administration of Philippine roadwork more broadly as Secretary of Commerce and Police and Governor-General, are examined in Chapter four.
21 A devoted polo player, Forbes also played on the Harvard football team and served for two years (1897–1898) as its head coach.
22 Forbes, "Journal" (February 1904), First Series, Vol. I, p. 2.
23 Forbes, "Journal" First Series, Vol. I, p. 307, n. 49. Underscores original.
24 In Hines, *Burnham of Chicago*, p. 178.
25 Gerard Lico, "Spatial Regulation as Prophylaxis: Urban Hygiene and Colonial Architecture in the Age of American Imperialism in the Philippines," in Gerard Lico and Lorelei D.C. De Vianna (eds.), *Regulating Colonial Spaces (1565–1944): A Collection* of Laws, *Decrees, Proclamations, Ordinances, Orders, and Directives on Architecture and the Built Environment During the Colonial Eras in the Philippines* (Manila: National Commission for Culture and the Arts, 2017), pp. 107–152, p. 108, p. 111.
26 "The Conquest of Bubonic Plague in the Philippines" *National Geographic Magazine* XIV (May 1903), pp. 185–195; Warwick Anderson, *Colonial Pathologies: American Tropical Medicine, Race, and Hygiene in the Philippines* (Durham, NC: Duke University Press, 2006), pp. 45–73.
27 Lico, "Spatial Regulation as Prophylaxis," p. 111.
28 Lico and De Vianna, *Regulating Colonial Spaces*, pp. 153–189.
29 Lico, "Spatial Regulation as Prophylaxis," p. 128.
30 Don Mitchell, *The Right to the City: Social Justice and the Fight for Public Space* (New York: Guilford Press, 2003), p. 187.
31 Frequent slide-outs after heavy rains meant the road was far from a one-time cost. McCoy pegs construction costs for the Benguet Road at "an astronomical 4.1 million [pesos]" from 1902 to 1906, "even more than those for Manila's massive new port." Alfred McCoy, *Policing America's Empire: The United States, the Philippines, and the Rise of the Surveillance State* (Madison, WI: University of Wisconsin Press, 2009), p. 253. By Interior Secretary Dean Worcester's admission in *The Philippines Past and Present*, a work of unparalleled apologia for the American regime, more than US$ 2.75 million had been spent on construction and maintenance of the Benguet Road by 1913 (or more than US$ 77 million in 2021 currency). Worcester, *Philippines Past and Present*, p. 257.
32 Anderson, *Colonial Pathologies*, pp. 142–147.
33 Forbes, "Journal" (entry 12/08/1904), First Series, Vol. I, p. 123.
34 Forbes to Burnham, 29 August 1904. On the norm of American colonial elite, after British imperial norms in South Asia, relying on Chinese servants to run their households during this period, despite policies of Chinese exclusion, see Julia Martínez and Claire Lowrie, "Transcolonial Influences on Everyday American Imperialism: The Politics of Chinese Domestic Servants in the Philippines," *Pacific Historical Review* 81 (2012): 511–536.
35 *Chicago Daily Tribune*, "Plan Queen City for the Far East," September 18, 1904, p. 1. Daniel H. Burnham Collection, Box 56, FF 19. Ryerson and Burnham Archives, Art Institute of Chicago.
36 Burnham (Daniel H.) to Burnham (Margaret S.), December 7, 1904. Daniel H. Burnham Collection, Box 25, FF 13. Ryerson and Burnham Archives, Art Institute of Chicago.
37 Burnham and Anderson, "Report on Proposed Improvements at Manila."
38 Burnham in Charles Moore, *Daniel H. Burnham: Architect, Planner of Cities* (Boston, MA: Houghton Mifflin, 1921), p. 238.
39 Ibid. Burnham's Philippine visit was recorded in broad strokes in Burnham's diary, a terse document that offered, in the words of Burnham's authorized biographer, "quick glimpses of the life, both social and professional, of a

surpassingly busy man, and also reveal his method of work" (Moore, *Daniel H. Burnham*, p. 246). Work perhaps, dinner certainly. As Moore continues, "It was Mr. Burnham's practice each day to note briefly in his diary the people he met, the topics discussed, and any circumstances the date of which might be important. This he did for his own protection against people who either from bad memory or design might make misstatements as to past occurrences." Indeed, Burnham's most consistent concern in his Philippine diaries appears to be noting down his dinner arrangements. Yet strikingly, despite Burnham's otherwise careful record of his companions at the events, he mentions no Filipinos by name in the diaries; on a few occasions, Burnham observes that there were "also several Filipinos" in attendance but they remain, like the dug-out paddlers, unnamed. Curiously, the diary of Burnham's Philippine visit (and trans-Pacific travels) was published (or excerpted) within Moore's (1921) biography, these travel journals were not available among Burnham's diary pages from the Burnham Papers microfilms at the Ryerson and Burnham Libraries. As another researcher noted decades ago, although Moore had access to the travel journal in 1920, evidently it was not included in the collection of diaries and papers in the Burnham family's gift to the Art Institute of Chicago (Hines, "Imperial Façade," p. 42, n. 14). One possible source of embarrassment in the uncensored diaries may have been the expression of Burnham's views on race, which can be pieced together from other traces, such as an October 1905 letter from future US Secretary of War Jacob Dickinson thanking Burnham for the copy of a book, entitled *Anglo-Saxon Superiority*, that the architect had furnished for him. Dickinson to Burnham, October 24, 1905. Daniel H. Burnham Collection, Box 1, FF 21. Ryerson and Burnham Archives, Art Institute of Chicago. All references from the diary in this section are to "The Diary Records" in Moore *Daniel H. Burnham*, 234–245, unless indicated.

40 Forbes, "Journal" (entry January 1, 1905), First Series, Vol. I, p. 129; Burnham in Moore, *Daniel H. Burnham*, p. 240.

41 Forbes, "Journal" (entries December 22 and December 26, 1904), First Series, Vol. I, pp. 127–128.

42 Forbes, "Journal" (entry December 26, 1904), First Series, Vol. I, p. 128.

43 Like Manila, Baguio's development under U.S. rule was by no means solely an aesthetic endeavor. On the town's emergence as a regional market and administrative node at the center of what was to become the Mountain Province, see Gerard A. Finin, *The Making of the Igorot: Contours of Cordillera Consciousness* (Quezon City, Philippines: Ateneo de Manila University Press, 2005), pp. 41–76; McKenna, *American Imperial Pastoral*, pp. 111–141.

44 In Moore, *Daniel H. Burnham*, p. 240.

45 Forbes, "Journal" (entry January 8, 1905), First Series, Vol. I, p. 131.

46 These sketches, given directly to Forbes, Governor Wright, and city and US Army engineers, did not appear in the publication of Burnham's official report but are included among the list of sketches with brief descriptions in Daniel H. Burnham, *Plans of Manila, P.I.* Official plans furnished by D.H. Burnham, June 28, 1905. Daniel H. Burnham Collection, Box 56, FF 21. Ryerson and Burnham Archives, Art Institute of Chicago. The whereabouts of the sketches themselves are unknown to me.

47 Moore, *Burnham of Chicago*, p. 245.

48 Burnham and Anderson, "Report on Proposed Improvements at Manila."

49 Daniel H. Burnham and Pierce Anderson, "Preliminary Plan of Baguio Province of Benguet P.I.," June 27, 1905. Daniel H. Burnham Collection, Box 56, FF 3. Ryerson and Burnham Archives, Art Institute of Chicago; Daniel H. Burnham and Pierce Anderson, "Report on the proposed plan for the City of Baguio, Province of Benguet, P.I.," submitted October 3, 1905. RG 350/Entry 5/

Box 534. National Archives and Records Administration (NARA), College Park, Maryland.

50 Burnham and Anderson, "Report on Proposed Improvements at Manila," p. 635.

51 Morley, "Modern Urban Designing," p. 13.

52 Morley, "Modern Urban Designing," pp. 10–21; Cristina Evangelista Torres, *The Americanization of Manila, 1898–1921* (Quezon City: University of the Philippines Press, 2010).

53 Burnham and Anderson, "Report on Proposed Improvements at Manila," p. 627.

54 Ibid.

55 Ibid., p. 631

56 Ibid., p. 629.

57 Ibid., p. 630. Burnham would not get to see Intramuros before U.S. Army engineers had already begun to blast through its walls in an effort, justified as a public health concern, to better ventilate and reduce congestion in the urban center. "It is a very small fraction of a very big wall," Forbes explained to Burnham, "and in the less attractive part of the city." Forbes to Burnham, August 29, 1904. Burnham disagreed, observing the old walls' "singular historical and archaeological influence and imposing appearance of monumental value," and noting that, as "obstacles to the free circulation of air, their moderate height compared with adjacent buildings seems to make them comparatively unobjectionable." But in the interest of efficiency he did not object to new street openings through the Intramuros provided they were cut through the projecting bastions at the original angles of the structure. Burnham and Anderson, "Report on Proposed Improvements at Manila," p. 627.

58 Burnham and Anderson, "Report on Proposed Improvements at Manila," p. 631.

59 Ibid., p. 628.

60 For architectural and planning history perspectives on the Burnham plans, see Hines, *Burnham of Chicago*, pp. 197–216; Lico, "Building the Imperial Imagination"; Brody, *Visualizing American Empire*, pp. 140–163; Torres, *Americanization of Manila*; Vernon, "Daniel Hudson Burnham"; Morley, "Modern Urban Designing," pp. 10–21;

61 Burnham and Anderson, "Report on Proposed Improvements at Manila," p. 628.

62 Ibid., p. 632.

63 Ibid., p. 631.

64 Ibid., p. 633.

65 Ibid., p. 628.

66 Ibid.

67 Ibid., p. 629.

68 Ibid.

69 Ibid., p. 633.

70 Cf. Morley, "Modern Urban Designing," pp. 13–16.

71 Burnham and Anderson, "Report on Proposed Improvements at Manila," p. 633.

72 Ibid.

73 Ibid.; cf. G.W. Browne, *The Pearl of the Orient* (Boston, MA: Dana Estes and Co, 1900).

74 Burnham and Anderson, "Preliminary Plan of Baguio," p. 1.

75 Ibid.

76 Ibid., p. 2.

77 Forbes, "Journal" (entry September 17, 1904), First Series, Vol. I, p. 69.

78 Forbes to von Meyer, October 12, 1909. W. Cameron Forbes Papers, MS Am 1366.1, Confidential Letter Book No. 1, Houghton Library.
79 Brody, *Visualizing American Empire*, p. 154.
80 As McKenna has shown through the case of *Mateo Cariño v. the Insular Government*, Baguio's government reservation was itself legally contested real estate. McKenna, *American Imperial Pastoral*, pp. 82–94.
81 "We also want some building to give poor people a chance to use what we've opened up," he added vaguely, perhaps perfunctorily. Forbes, "Journal" (entry January 8, 1905), First Series, Vol. I, pp. 130–131.
82 In Brody, *Visualizing American Empire*, p. 154.
83 Anderson's preliminary sketch of Topside can be seen (in grayscale) in Scott Kirsch, "Aesthetic Regime Change: The Burnham Plans and US Landscape Imperialism in the Philippines," *Philippine Studies: Historical and Ethnographic Viewpoints* 65 (2017): 315–356, and in the original color in W. Cameron Forbes Papers, Additional Papers (MS Am 1192.13, p. 128), Houghton Library, Harvard University.
84 Forbes, "Journal" (entry February 3, 1905), First Series, Vol. I, p. 143.
85 Forbes, "Journal" (entry May 1, 1905), First Series, Vol. I, p. 201 (emphasis added). The pledge to maintain the Benguet Road occurred alongside additional state investments in the production of space, including a half-million U.S. dollars set aside for new wharves in Manila. On American efforts to transform land and labor at the Port of Manila at this time, see Mike B. Hawkins, "From Colonial Cargo to Global Containers: An Episodic Historical Geography of Manila's Waterfront," unpublished PhD. dissertation, Department of Geography, University of North Carolina at Chapel Hill (2022), pp. 25–85.
86 Forbes, "Journal" (entry June 26, 1905), First Series, Vol. I, p. 240.
87 Although the plan proposed in the long term to redirect civic and commercial activities to the new esplanade and unbuilt quarters of Manila, it offered the U.S. regime an organic framework for modifying Manila's street system, which could be undertaken without upsetting local class interests. Hence, for Burnham and Anderson, "The first step consists in establishing the new street lines and purchasing such real estate as can be acquired without damage to property interests." In turn, streets could be purchased in piecemeal fashion, allowing that the "completion of the work by removal of the remaining buildings" could be left "for half a century if necessary until such time as public safety calls for their demolition, or until the owners on their own initiative decide to rebuild in accordance with the altered street lines." Burnham and Anderson, "Report on Proposed Improvements at Manila," p. 630.
88 Forbes, "Journal" (entry September 5, 1905), First Series, Vol. I, p. 307.
89 Forbes, "Journal" (entry May 26, 1906), First Series, Vol. II, pp. 16–17.
90 For architectural historian Thomas Hines, in ranking Parsons's work among the highpoints of American modernism, "convincing Parsons to execute the Philippines work was indeed one of Burnham's most fortunate accomplishments." Hines, *Burnham of Chicago*, p. 211. See also Thomas S. Hines, "Modernism in the Philippines: The Forgotten Architecture of William E. Parsons," *Journal of the Society of Architectural Historians* 32 (1973): 316–326; Brody, *Visualizing American Empire*, pp. 154–163; Morley, "Modern Urban Designing." Architectural historian Diana Jean Sandoval Martinez describes a different high point of U.S. colonial architecture in buildings initiated under consulting architect Ralph Harrington Doane (1916–1918) which featured a sparkling white marble and cement aggregate derived from local materials in place of Parsons' more austere uses of reinforced concrete. Apparently both visually pleasing and politically acceptable, the new landscape surfaces were relatively short-lived (and do not remain visible); after heavy destruction from

two World War II invasions, cheaper building methods and regular cement were used in their reconstruction. Martinez, "Concrete Colonialism: Architecture, Infrastructure, Urbanism and the American Colonization of the Philippines," Ph.D. dissertation, Columbia University (2017). Available at: https://academiccommons.columbia.edu/doi/10.7916/D88P6BXX

91 Parsons lauded the "cooperation of American and Filipino draftsmen in the drafting room" in forming "an economical organization," but emphasized that the "need of additional American assistants … is felt in the larger and more important problems." The Office also relied on clerical work, stenography, engineering, and surveys from the Bureau of Public Works, under Forbes's administration as Secretary of Commerce and Police. William E. Parsons, "Annual Report of the Consulting Architect for period extending from November 17th, 1905 to June 30th, 1906," June 30, 1906, p. 3. Daniel H. Burnham Collection, Box 56, FF 22. Ryerson and Burnham Archives, Art Institute of Chicago.

92 Parsons, "Annual Report of the Consulting Architect," p. 13.

93 As Mike Hawkins observes, the "newly-arrived American colonial regime dynamited, quarried, dredged, and dumped countless tons of rock, stone, concrete, and seabed at a breath-taking pace. Yet, the unfathomable scale and perhaps, the government's own seemingly impossible accounting of each ton of stone and cubic yard of seafloor used during construction was the point, lending a sense of permanence, stability, and carefully-accounted rationale to the regime." Hawkins, "From Colonial Cargo to Global Containers," p. 29.

94 "Secretary Dickinson to Lay Corner Stone of New Hotel Which Will Be Most Complete in Orient," *The Manila Times* August 29, 1910, p. 1.

95 Forbes, "Journal" (entry May 5, 1906), First Series, Vol. II, p. 5.

96 Ibid.

97 W. Cameron Forbes Papers, Houghton Library, Harvard University; Dean C. Worcester papers, 1863–1915, Worcester Philippine Collection, University of Michigan Special Collections Library.

98 Worcester, *Philippines Past and Present*, Vol. 1, p. 262.

99 The Country Club Corporation was itself owned in stock by subscribers (members), with Forbes purchasing purchase ten shares (at $50)—no other subscribers purchased more than three—to "help the club get started." Worcester, Philippines Past and Present, Vol. 1, p. 262.

100 "Our White Elephant," *Philippines Free Press* August 19, 1911, p. 1; cf. "Las Delicias del Veraneo," *El Renacimiento* March 13, 1908, p. 3.

101 Still, Forbes complained, Baguio remained "only half a success, that is to say, the journey up is still slow and carries enough discomfort to make it a hardship for weak or tired people, and half the people who go up there don't like it, probably, I think, because there is nothing to do, no society, no games, no drives, no men folk about and no very good place to stay." Forbes to Burnham, May 14, 1905. In W. Cameron Forbes, *Notes on Early History of Baguio* (Manila: Manila Daily Bulletin, 1933), p. 14.

102 Parsons, "Annual Report of the Consulting Architect," p. 8.

103 Ibid. By 1910, 295 town lots had been sold, but only ten residences built (in addition to scores of rentable cottages), raising concerns that Baguio had become chiefly a speculative venture but here too construction proceeded more rapidly in the 1910s, buoyed by the establishment of waterworks, waste systems, and electric light and power production. McKenna, *American Imperial Pastoral*, pp. 142–173.

104 Parsons, "Annual Report of the Consulting Architect," p. 8.

105 Ibid., p. 9.
106 Worcester, *Philippines Past and Present*, Vol. 1, p. 264.
107 Ibid.
108 Ibid.
109 By 1911, McKenna notes, some 600 government workers—250 Americans and 350 Filipinos—were making the trip to Baguio for the summer session. McKenna, *American Imperial Pastoral*, p. 142.
110 Rodney J. Sullivan, *Exemplar of Americanism: The Philippine Career of Dean C. Worcester*, Michigan Papers on South and Southeast Asia 36 (Ann Arbor, MI: Center for South and Southeast Asian Studies, University of Michigan, 1991), pp. 146–148.
111 Harrison (1922) in ibid., p. 147.
112 Forbes to Dickinson, June 8, 1910. W. Cameron Forbes Papers, MS Am 1366.1, Confidential Letter Book No. 1, Houghton Library.
113 The "tribes" were enumerated and ranked by Forbes in order of "their aptitude for the making of soldiers," with Ifugaos, Bontoc Igorots, and Kalingas leading the field, followed by groups—Lepanto Igorots, Apayos, Benguet Igorots, and Amburayan Igorots—deemed "comparatively peaceful people" and "much better as cargadores or laborers and less dependable as warriors." Forbes to Pershing, March 2, 1912. W. Cameron Forbes Papers, MS Am 1366.1, Confidential Letter Book No. 1, Houghton Library.
114 Ibid. If Pershing responded to the proposal it was not found among Forbes's letter books.
115 W. Cameron Forbes, Philippine Data, Personal, Vol. 1, pp. 128–129. W. Cameron Forbes Papers, MS Am 1192.4, Houghton Library.
116 Forbes, "Journal," First Series, Vol. II, p. 167, note (1930).
117 Forbes, "Journal" (entry April 28, 1907), First Series, Vol. II, p. 217; Topside Guestbook, by various hands, 1906–1912, W. Cameron Forbes Papers, MS 1365.4, Houghton Library.
118 Forbes, "Journal" (entry April 28, 1907), First Series, Vol. II, p. 217.
119 In this landscape of aesthetic delight, McKenna observes, Americans cast the Igorots "as curiosities and landscape features on par with sensational views of non-human nature." McKenna, *American Imperial Pastoral*, p. 145.
120 Lefebvre, *Production of Space*, p. 189.

4 Road

W. Cameron Forbes, Philippine Roadwork, and the Production of Space

Road Bible

"I have written a letter which I am going to have translated into all the dialects and sent to every municipal president, provincial governor, engineer, and road foreman in the Islands," Cameron Forbes noted for his journal on 30 May 1908, "on the necessity of roads and what has been done."[1] The 13-page circular, he ventured, would become the "road bible" of the Philippines.[2] Articulating a politics of material development while positioning "good roads" as a technical matter *beyond* politics, road (and bridge) building would also provide Forbes with a signature issue on his journey to the Governor's mansion at Malacañang. Printed and distributed two weeks later, the extraordinary letter emphasized the "very lamentable condition" of Philippine roads, and the necessity of both expanding and better maintaining the roads if "these Islands" were to "expect any real measure of prosperity."[3] Although, like Secretary of War Taft, Forbes largely blamed Filipino municipalities for declining road conditions,[4] his letter did not dwell in the past. The Insular, provincial, and municipal authorities had all underfunded the maintenance demands of roadwork, he acknowledged, even as the government was expanding and reconstructing the road network, resulting at times in a net loss of road mileage, as roads slid into disrepair. What was needed, Forbes argued, was "a complete change in the system of road construction and maintenance—one which will provide not only for the construction and repair of all necessary roads but for their continuing maintenance thereafter."[5]

System, perhaps. Ideology, certainly. As Forbes put it, in what may have seemed to both American and Filipino officials and road foremen as a tedious a lesson in political economy:

> No matter how rich, fertile, and productive land may be, the owners and inhabitants of it can only realize a fraction of its value if there is no market for its products. They are thrown back into a primitive state of living where each person instead of producing the things which he can produce most economically and selling them, and with the money

DOI: 10.4324/9780429344350-5

> buying things which others can produce economically, will produce for himself a poor class of cloth, a poor class of food, a poor class of shelter, and practically none of those other things which people manufacture and sell and which tend to make modern life agreeable. The construction of a road immediately puts the vigor of life into the agriculture and industry of the region which it opens.[6]

"Good roads" were keen indicators of future success, Forbes insisted, for "the better the road the more prosperous the province." Smooth, level roads, well drained, properly crowned, surfaced, and metaled, would allow a carabao to haul four or five times as much produce on carts than "if he had to drag the wheels through mud, in and out of holes, down into rivers, and up steep banks on the other side." Name-dropping Napoleon, Forbes insisted that the lessons of history were clear: "all well-governed countries pay great attention to their roads, and in fact it can be taken almost as an axiom that the merit of a government and the degree of efficiency of it its administration can be measured by the condition of its roads."[7] And yet the impacts of roadbuilding, from the work required to produce and maintain the roads to the work that roads performed, were socially, politically, and geographically complex, tending to provide the greatest benefit to large landholders while rendering a new terrain on which the scalar relations of power among the Insular and provincial governments could be recast. The roadbuilding *terrain* was no mere political metaphor. Opening new pathways for transportation and commerce, better roads and trails also extended the geographical reach of the state, from the taxman to the Constabulary inspector, into heretofore more isolated places.

What Forbes called his "road movement" or "road campaign" was well underway by the time he penned his good roads gospel, including efforts to formally re-instigate the Spanish corvée, compelling five days of roadwork per year (later upped to ten) from Filipinos unable to afford a regressive *cedula* (personal registration) tax. The burdens of compulsory roadwork would fall disproportionately on the poor rural peasantry, particularly among the mountain peoples of northern Luzon, where the Philippine Constabulary claimed the authority to conscript road workers, and in the Southern archipelago, under U.S. Army rule. Forbes's road circular, encompassing legislative history, political philosophy, and intergovernmental strategies—stands as a compelling artifact of the imperial moment, a coherent ideological code for the recasting Philippine transportation infrastructure under U.S. colonial rule. Eyeing an earlier imperial moment, Henri Lefebvre described the "Roman road, whether civil or military," as a key element of relational spatial practices, materially linking "the *urbs* to the countryside over which it exercises dominion. The road allows the city, as people and as Senate, to assert its political centrality at the core of the *orbis terrarium*."[8] And if, as Lefebvre argued, an ideology only "achieves consistency by intervening in social space and its production, and by thus taking on body therein," then that is precisely what Forbes hoped to achieve in his road campaign, constituting what he saw as a pivotal dimension of the American (re)production of colonial state space in the Philippines.[9]

From 1907 to 1913, the Insular government boasted the construction of 1,000 miles of "first class roads."[10] While each new mile could be touted as evidence of material progress under U.S. sovereignty, each was also the product of backbreaking and often dangerous human toil, much of it coerced under the auspices of the Insular government, U.S. Army, Philippine Constabulary, and provincial and municipal authorities. This chapter explores the project of Philippine roadwork under the Taft-Forbes regime as a spatial strategy of empire that presented new problems of labor and geography to be overcome, revealing, in turn, new contradictions of liberal empire. It traces Forbes's efforts to seek solutions to the challenges of road construction through the revival of the corvée, experiments with prison labor, and constitutional strategies for ensuring consistent funding and labor. Forbes also introduced a rigorous road *inspection* regime which sought to punish local politicians who did not adequately maintain the roads and his own "caminero system" of road maintenance. But if Forbes was the leading edge of a coercive colonial state attempting to realize sweeping transformations in the Philippine landscape and space-economy, then the forms that state power took were geographically complex, as persistent demands for roadwork also animated a changing scalar politics of labor procurement, and provincial elites leveraged their own positions in networks of power as providers of labor and builders of roads.

In October 1907, Secretary of War Taft returned to Manila for the inauguration of the Philippine Assembly, extolling the moment as "another step in the enlargement of popular self-government in these Islands."[11] With the American-dominated Philippine Commission now acting as an "upper house" to the new Assembly, and with Taft's tacit support, Forbes would pursue legislative strategies aimed at entrenching his ostensibly perfect systems of roadbuilding, maintenance, and inspection in the archipelagic landscape through a permanent appropriation bill. Yet Forbes remained surprised, or conspicuously *not* surprised, by the shifting alliances that constituted Filipino nationalist politics. Reflecting in his journal on the passage of a 1908 appropriation bill which, though not a "permanent appropriation," placed two million pesos at his discretion for road and irrigation works, Forbes observed:

> The Progressistas or so called government party or pro-American party, that have done everything they possibly could do to embarrass us, fought this provision. The Nationalists who are the so-called anti-American party, and who have done anything they possibly could to assist the government … crowded it through. Osmeña brought the bill up twenty minutes before the regular session ended and crowded it

> through like a whirlwind, and although it leaves out a number of things we ought to have … I am reasonably content.[12]

The alternative to good roads, as Forbes put it starkly in his road bible, was "remaining in a condition of hopeless poverty."[13] But Americans' reliance on local allies to achieve their agenda, like Sergio Osmeña, the former Cebu Governor and first Speaker of the National Assembly, also underscored the complex relations through which colonial state spaces were reproduced.[14]

In the next section, we turn to problems of labor and landscape, along with corruption and graft, as American engineers oversaw the construction of the Benguet Road to Baguio, ultimately joining Forbes in his first encounter with Philippine roadwork in 1904 shortly after his appointment to lead the Department of Commerce and Police (which included the of Bureau of Public Works). Eyeing the development of the caminero system and other road policies under Forbes's leadership, the chapter turns to legislative and practical dimensions of the roads program as both a productive and disciplinary governmental project. Forbes's evolving road campaign, I suggest, was less a road bible than a kind of "moral road atlas" around which regimes of inspection, equating good roads with good government, could be organized.

Encounters with Roadwork

Nearly from the start of the occupation, Americans were intent on efforts to build and reconstruct Philippine roads. The advantages of improved roads in facilitating troop movements appeared as obvious to a U.S. Army engaged in counter-insurgent warfare, and the Army claimed, by 1900, to have directed the construction or repair of more than 1,000 miles of road.[15] A civil regime committed to modernization, revenue production, boosting exports, and the accessibility of its proposed summer capital and resort at Baguio, in turn embraced "good roads" as an engine of development. But the labor of colonial bureaucrats and military engineers meant little without the manpower to build the roads themselves. In the first years of the decade, the demands for labor, the living bodies required to swing a pickaxe, run a drill, manage a crew, or carry supplies up the trail, produced new challenges under conditions Americans understood as persistent labor shortages. In the wake of (and amid ongoing) warfare, hunger, cholera outbreaks, and pandemic animal disease, work on the roads, under an American boss likely to berate workers or worse, evidently held little appeal, even for a largely impoverished peasantry.[16]

For the Army and Insular state, one response to perceived labor shortages—which, along with seasonal rainfall, could grind American roads projects to a halt—was to take advantage of legal ambiguities (and the continuance of elements of the Spanish law) to claim the authority to impress workers when needed, including mountain peoples who had only been introduced in limited ways to a cash economy and were without recourse to money to avoid the work.[17] While, unlike the former Spanish corvée, American officials typically *paid* the impressed Filipino roadworkers, either in pesos or sacks of rice, the labor was nonetheless coerced, revealing fissures and contradictions in visions of a liberal—and liberatory—U.S. empire. Increasingly, however, as Americans drew elite Filipinos into collaborative positions in local and provincial governments, they relied on local *presidentes*, provincial officials, and labor brokers, interested in establishing or restoring their own advantageous arrangements, to supply road labor under different mechanisms which sometimes straddled the line between free and unfree labor.[18] These localized geopolitics of labor procurement thus reflected not only the contradictions of American liberal imperialism but also the distributed nature of U.S. colonial power in the archipelago, even in early moments of sovereignty in the Philippines.

The experiences of those drafted to the roads, not surprisingly, are scarcely recoverable from the colonial archive, though traces survive. Some bear witness to moments of catastrophe and horrible sadness, such as a note in Forbes's journal after meeting with individuals from Ifugao province in northern Luzon who had requested "that they not be sent out of their country for road work." Their pleas, Forbes recalled, "had reference to a very unfortunate affair, namely the construction of the Nueva Ecija-Nueva Viscaya Road … as this was the starting place from which they began to work."[19] Conscripted "to build a road in which their province had little interest," their wages had been stolen by a corrupt foreman. Even worse, the Ifugaos had been sent to an area with an active influenza epidemic, "and the returning laborers brough the infection back into the hills," with unspecified consequences.[20] Additional traces of the experience of road workers, and of life and death on the road, are revealed indirectly in the record of minutes, investigations, and reports generated in the context of conflicts and oversight within state power, as in the example of a lengthy interview with N.A. Holmes, then chief engineer of the Benguet Road, by members of the Philippine Commission, comprising Commissioners James Smith, Dean Worcester, Luke Wright, and Governor-General Taft, in a June 1903 meeting.[21]

The Benguet Road, the proposed 38 mile stretch of wagon road, criss-crossing the Bued River in its ascent to the budding American summer capital and mountain resort at Baguio, was the most technically ambitious—and politically tone-deaf—of the Insular government's early road-building efforts, as described in Chapter 3. Rising spectacularly in a series of steep switchbacks known as the Zig Zag (see Figure 4.1), laying the road into the Cordillera Central (where only horse or foot trails had previously existed) required immense resources of labor, funding, explosives, and logistics. While Baguio boosters Forbes and Worcester would insist that the road had been built to offer healthful alpine access to lowland Filipinos as quickly as possible, it was the aesthetic landscape—and colonial fantasy—of Baguio, that animated the project, and accounted for the urgency and

Figure 4.1 The "Zig Zag," Benguet Road (now Kennon Road).

Source: Photo by the author.

priority that it was persistently conferred from Insular government revenues in construction and repair costs. By the time of the Commission's June 1903 hearing with Holmes, the Benguet Road was plagued not only by a shortage of labor but also questionable routing, mismanagement, and graft. Progress had slowed to a crawl, with just two miles of wagon road and two additional miles of horse trail completed in the past year, and the Commissioners, exposed to criticism from both Filipino nationalists and American anti-imperialists, were beginning to lose patience.

In the face of Commissioners' criticism, Holmes's explanations focused equally on problems of labor and those posed by the terrain itself, including one stretch of four miles of nearly continuous rock in which nothing had yet been accomplished, and others wherein the rock was covered with a thin layer of soil so that "practically the entire road has to be cut through the rock."[22] An additional ten miles of the proposed roadway remained untouched. For the heavy "rock work," as distinguished from less demanding "dirt work," Holmes pointed out on a map where workers toiled some 200 feet above, chipping downward to exploit small irregularities in the loose and scaly rock bench grade (see Figure 4.2). Much of the rock work consisted of drilling holes into the rock and setting off charges of black powder; for those tasked with lighting the short fuses or disposing of unexploded

Figure 4.2 Road work in the Bued River Valley. In this unidentified surviving image of ledge work on the sheer mountainsides of the Bued River valley, we can make out the ladders extending upward from the lower left corner of the image to a high ledge or platform on which the figures of perhaps 20, likely conscripted workers can be seen chipping away at the side of the mountain under precarious conditions.

Source: Courtesy, National Library of the Philippines (image 03/887).

charges, or for many others vulnerable to the landslides that the explosions sometimes caused, the conditions were particularly dangerous.

Holmes was quick to volunteer for Commissioners a racialized calculus of key differences among the work crews. Asked about the costs of rock work, Holmes estimated one peso (worth 50 cents) per cubic yard excavated. But while the Filipino workers, Holmes attested, typically drilled from 15 to 30 inches per nine-hour workday (for three men operating a drill), American crews typically drilled from 6 to 12 feet per day, according to foremen's reports. Hence, American laborers had been hired at a much higher rate of two dollars per day plus subsistence (compared with 25 to 50 cents per day for Filipino workers), though this wage had turned out to be too high, Holmes conceded, and was reduced to $1.50 per day for American labor to remain more efficient than Filipino workmen, due largely to the extensive sick days Americans tended to require.[23] Holmes estimated 12 to 15 American patients in the camp hospital every day out of about 200 Americans, after which many were summarily dismissed to avoid paying subsistence costs before the men were able to return to work.

"The class of men you get are pretty hard drinkers?," asks Commissioner Luke Wright, then Secretary of Commerce and Police. Holmes waivers but, keen to deflect blame, does not sugarcoat his assessment of the American roadworkers:

> No, the thing against them is that they are generally the worst of characters, men whom the police are looking for or who have been thrown out of work in other parts of the Islands, and most of them are hard drinkers, and the consequence is that their system is depleted; and if they are thoroughly healthy they generally get into some bad scrape before long and have to be discharged.[24]

And yet it is the Filipino laborers, on whose efforts he overwhelmingly relied, that Holmes holds in special disdain:

> The Filipino is a remarkably poor workman, and after my previous experience with him I can see no possible use for the Filipino as a laborer. We have tried every possible method that all of us could think about to bring the Filipino up into some efficiency and have met with not a particle of success. He doesn't want to learn. In the first place, he doesn't want to work.[25]

Of course, it is hardly surprising that these workers were not keen to chip rock downhill from hundreds of feet above the trail, or to spend their days boring into rock to emplace explosives, even if they were being paid (albeit a fraction of their loutish American counterparts). Although Holmes does not seem to consider that the Filipino workers' slower pace might itself have been a form of resistance, his observations shed light on conditions of coerced or unfree labor on the road, as well as the scalar distribution of power embodied in the collaboration of local mayors (*presidentes*) under the Insular state:

> I can safely assure you that there is hardly a Filipino who comes up to work that is not practically forced to come there, and that is done through the *presidente*. The *presidente* will issue an edict that so many men in such a barrio must come out, and they come, and the only possible reason that any of them come is that they are sent there by that

> *presidente*. We have a few here and there in the nature of skilled workmen, like a mason, a carpenter, or a capataz, who will come in to work, but they are very few, and the majority of the workmen come because they are actually forced to do so.[26]

Holmes appears to hold his own under the Commissioners' scrutiny, at least until their interrogation turns to the existence of mineral claims and the operation of a working mine—the Copper King—by current and former road engineers (including Holmes himself). Sketchy stories of missing powder kegs used in mining operations, "borrowed" and later repaid to government stores, suggest that the Benguet Road project under Holmes did not suffer only from labor shortages and poor workmanship.

Shortly after the hearing, the Benguet Road was placed under the command of Major Lyman Kennon—with Holmes staying on as chief engineer—and it is Kennon, who rammed the project through to completion over the next two years, for whom the road is named today.[27] Of course, the designation of the spectacular mountain road after a single one of its makers—among the tens of thousands who labored on it in short stints—throws the contradictions of hagiographic individualism in naming practices into sharp relief. If Kennon did demonstrate special capacities in his efforts on the road, what stands out is his skill in maintaining a large labor force, which did not fall under 2,500 workers under his leadership.[28] Relying increasingly on labor agents, and the targeted recruitment of Ilocanos from northwest Luzon, Kennon and his workers blasted ahead at full bore along the route prescribed by his predecessors.[29] Baguio enthusiasts on the Commission, seasoned by disappointment and cost overruns, were delighted by the new pace of roadwork and reports of Kennon's apparent success.[30] While the Commission restricted itself, for a time, from spending on the development of the summer capital itself, which remained deeply unpopular among Filipino nationalists, it opened the Insular purse strings for Kennon, providing in its subsidies the strongest possible guarantee for the summer capital idea.

Whether the road should have been built at all, and built *where* it was built, were different questions. Forbes, much later, observed that "Major L.W.V. Kennon was an enthusiast and almost a martinet. … He was a man of great presence, tremendous force and drive and, I believe, not entirely sound in judgement."[31] But if Forbes had misgivings about Kennon at the time he did not express them, noting admiringly in his journal that "from Twin Peaks up the work is the Major's and he is justly proud of his achievement."[32] Forbes would remain a booster of the Benguet Road during his Philippine career but as he would later reflect on Kennon's work, taking the full benefit of hindsight:

> I came to feel afterward that he had done exactly the wrong thing about the Benguet road. He should have insisted upon the survey before going ahead; he should not have accepted orders to build from Twin Peaks

> to Baguio and got that through in spite of the fact that he had become convinced it was the wrong place to go; he should have provided a good base to operate from instead of the winding road from Pangasinan, crossing the same river twice and going about fifteen miles out of the way; and generally he should have comported himself as one would expect a very careful and competent civil engineer, jealous of his reputation, to do. Instead of that he said he was performing a soldier's duty, that he took his order to build from Twin Peaks as an order and without offering suggestions against it, built from Twin Peaks up.[33]

The mix of rock and dirt work that Holmes described in his deposition would still have been underway, along with road grading and the construction of culverts of reinforced concrete, when Forbes, freshly appointed as Secretary of Commerce and Police, made his first visit to the road in September 1904. The journals reveal Forbes's excitement with the initial encounter, finding the road from Dagupan to Twin Peaks "as interesting as possible, thickly settled on both sides and the street crowded with people and traffic, much bound for Twin Peaks with stuff for the Benguet road. This is carried in two wheeled carts drawn by plodding oxen or carabaos, which go one mile an hour when they are in a hurry and lie down when they're not or when they can get into water."[34] Further up the road, "Huge lines of bullock carts go struggling through the mud," with "thousands of pounds taken on backs of Igorots, men, women, and children" to supply six to eight camps established along the road.[35] Like Holmes and Kennon, Forbes would view the workers themselves largely through lenses of race and nation. His observations also reveal the cosmopolitan character of a labor force that included a diversity of Filipino workers—including Tagalogs, Igorots, Pampangans, and Ilocanos—as well as Americans and "negroes" (listed separately), Chinese gang labor, Japanese, Spanish, Mexican, and Cuban laborers, carpenters, and masons, among others.[36] Not yet practiced in the Insular imperial discourse, which Forbes would internalize, of praising Filipino labor enough to encourage capital investment while denigrating it enough to shift blame for a lack of development, Forbes—likely echoing the statements of Kennon or his foremen—notes that the "Filipinos are the worst" of the workers, "afraid of heights and the rolling rocks, etc., and justly," he adds, for "few days pass without casualty, and up and down the huge canon can be heard the sound of the pick, the rattle of the rocks bounding down the mountain, then a loud splash in the river, the booming of dynamite and powder as the great blasts go, and strange cries of men."[37] In the same entry, Forbes describes a worksite known as "The Devil's Slide" due to the number of men killed on it; earlier that month, he notes, after celebrating a new allotment of $375,000 for Benguet road construction, that six men were killed on the road "a day or so ago from a landslide after a blast."[38]

Forbes would visit the Benguet Road at least five times during his first seven months in the Philippines, perhaps experiencing a gratifying sense of progress as Kennon and his crews chiseled and blasted their way toward the road's ostensible completion while gaining curious new glimpses of colonial life.[39] Describing Kennon's headquarters at Camp Four, he observes: "Inside the house were two delightful Filipina girls, sweet and friendly. One about thirteen, an orphan, Rosa, had been 'given' to Major Kennon, who makes her do a little housework, and the other, Maria, and her husband live and care for him without being willing to accept pay."[40] It is an odd phrasing, *without being willing to accept pay*, suggesting that Forbes did not know what to make of the arrangement, though relations of debt peonage were then relatively common in parts of the archipelago. The coterie of Rosa, Maria and her husband keeping house for Kennon at least captured his attention. But there were also clear takeaways for Forbes from his encounters with Kennon and Philippine roadwork. Among other Commissioners, Forbes was stung by criticism of the project's colonial elitism, and while he would continue to leverage his position in support of the Benguet road (and soon, Baguio construction), he would also intensify efforts to make "good roads," and the politics of material development, into more widely distributed phenomena in the archipelago. He also recognized that ensuring a large and steady supply of labor would be critical to his ambitious roadbuilding agenda. After Kennon's achievement, Forbes would aggressively pursue a range of strategies meant to meet those demands, transforming, at the same time, the nature of Philippine labor.

El Caminero

Known, perhaps apocryphally, as *El Caminero* (the roadman), Forbes would help to mobilize the Insular government around road construction as a self-consciously modern development policy, yet often in ways that threw the contradictions of his liberal imperialism into sharp relief. If, for the engineer Holmes, the desired solution to the labor problem on the Benguet Road had included sub-contracting Chinese labor gangs, then, for Forbes, a policy of expanded roadbuilding throughout the archipelago would produce more voracious labor demands and called for more systematic—and politically acceptable—solutions, including experimentation with different forms of coerced labor. The Boston Brahmin's diffuse bureaucratic portfolio, which, in addition to roadbuilding and public works included the administration of ports and harbors, railroads, light-house service, post office, business franchising, and prisons, allowed him to pursue a range of strategies to address the labor problem. Forbes drew on the contemporary discourse of "good roads" in North America, where the deployment of convict labor was becoming a key element in the roads movement, and the chain gang, composed primarily of African-Americans imprisoned for minor crimes and deployed to the roads, was emerging as the dominant penal institution in the American South.[41] That this brutal regime of convict labor could be cast as a progressive enterprise, modernizing the road network while offering supposedly healthful outdoor work for prisoners under the benign control of the state, made it an appealing ideology for road builders as well as rural landholders, and helps to explain Forbes's exuberance over the possibility of utilizing prisoners for Philippine roadwork. By March 1905, he had begun to make use of the prisons as another source of unfree workers, and the circulation of prisoners to road sites across the archipelago had become something of a pet project. As Forbes tallied their movements in one journal entry:

> Have been rushing off more prisoners... two hundred and fifty to Albay, which makes five hundred on that road and the full complement. General Wood wrote for two hundred and fifty more, which he'll now use on his railroad which he is going to build from Overton to Marahui. This makes five hundred in Mindanao, five hundred in Albay, two hundred and ten in Paragua, 150 in Malahi island, or in all 1360 moved.[42]

But the numbers would not add up to satisfy the labor demands for roads and other public works that Forbes hoped to produce across the archipelago. By cultivating collaborators on the provincial boards, however, Forbes would seek to secure more durable solutions to the problems of funding and labor for roadwork, and to develop what he saw as a perfect, "permanent" system of road construction and maintenance.[43]

As historian Justin F. Jackson has argued, forced labor on the roads "remained a legitimate policy option" in the specially administered zones of the Cordillera and Southern archipelago under the administration, respectively, of Philippine Constabulary and U.S. Army officers.[44] These norms for the Special Provinces were solidified in a 1905 Philippine Commission law requiring ten days of roadwork annually from males aged 18 to 60, or to provide a substitute; the resulting expansion of road and trail networks, along with new "Igorot Exchange" markets in the Cordillera provinces (which would be combined as Mountain Province in 1908), thus rested significantly on coerced labor.[45] Indeed, to the chagrin of liberal imperialist friends like Harvard President Charles Eliot, who intimated to Forbes his belief that the adoption of the "Spanish practice of forced labor on the roads" was a bridge too far, Forbes made accommodation with compulsory labor, as he had earlier with brutal policies of reconcentration, as a necessary facet of U.S. empire, and a means of accomplishing his ambitious roadbuilding agenda.[46] Forbes initiated efforts to reintroduce the corvée throughout the archipelago the following year in Philippine Commission Act No. 1511, which enabled provinces and municipalities (except Manila) to compel five days of road work annually from all able-bodied men or to demand a regressive tax "in default of the labor the commuted value thereof as fixed by the provincial board."[47] But the road law, the enforcement of which was optional for provinces and municipalities, "was not adopted by any province."[48] Hence, according to the Manila-based *Far-Eastern Review*, while Secretary Forbes had "endeavored to interest the natives in this great work," and some progress had been made, a "lack of initiative on the part of the Filipinos continues to obstruct the program."[49]

Whatever the basis of Filipino obstruction, Americans responded, in what had emerged as a key option in the Insular imperialist playbook, by ceding "large portions of the state to powerful Filipinos."[50] In 1907, Act No. 1652 authorized provincial boards to double the regressive cedula tax; the value of the cedula, for those unable to pay, could be "worked out on the roads."[51] While also ostensibly optional, the act differed from earlier road law in that the revenues raised were specified for road (and bridge) construction and maintenance at the discretion of the provincial board, rather than the municipalities, reflecting a shrewd scalar politics. The provincial boards were, from 1907, two-thirds selected by Filipinos, offering significant structural resources for those able to gain access to power at the provincial level.[52] To further entice provincial support for the road campaign, Commission Act No. 1695 provided an additional appropriation, authorizing fully 10% of internal revenue taxes to be divvied up for spending on roadwork. The funds were to be distributed proportionally (by population) but only among those provincial boards that had *accepted* the double cedula; appropriations for those provinces that did not accept the double cedula would be added to the pool of funds available for those that did. After the first year, all but four provinces "had the courage to accept the law doubling the cedula tax," as the *Far-Eastern Review* put it, "and thus enabled themselves to participate in the two million pesos of insular money made available for road purposes." Along with these funds, the added revenue would amount to 1.5 million pesos, bringing annual road and bridge funding to about four million pesos—some five times the amount spent in previous years. The impacts of these moves would leave an immediate and substantial impression on the Philippine landscape: in 1907 alone, 32 bridges and 276 reinforced concrete culverts were constructed along with five new steel bridges "at the expense of provinces active in this work."[53] Forbes's road campaign also garnered praise from Taft, who wrote personally, while enroute to Europe on the Trans-Siberian Railroad, to praise the accomplishments of Forbes's department, emphasizing "the movements that have been made toward an improvement of the roads, and the introduction of a satisfactory system with respect to them, the rule adopted for permanent improvement and construction."[54] Forbes's efforts were leaving an impression.

The 1908 "road bible" circular thus arrived at a propitious moment. Complaining that, in turning over maintenance of roads to the provinces and municipalities, "no definite system or rules for maintenance and construction" had been established, Forbes tacitly acknowledged that the expansion of roads had outpaced maintenance, leaving crucial work in the first years after construction unsupported.[55] As a result, bridges and culverts had been left exposed to heavy rains, un-graveled, and "not cared for"; many would rapidly deteriorate. Old Spanish roads had been allowed to fall into disrepair and were frequently impassible during rainy seasons, choking commerce. Whole provinces, Forbes insisted, had retrogressed. But the *caminero* system would "solve the problem of maintenance" through what Forbes expressed as a perfect scheme of maintenance and inspection. Under the caminero plan, adopted by the provincial boards under a resolution tied to the double cedula provision, first-class roads were divided into sections under a *capataz*, then further divided into subsections—averaging two kilometers of roadway (one during the rainy season)—that could each be maintained properly by a single caminero "who is responsible for the good conditions of that subsection."[56] Linking the Insular and provincial levels of governance to the everyday work of camineros, as well as the materials with which the roads were to be maintained, each caminero was:

> provided with a badge to indicate his office, giving the name of the province, the name of the road, and the number of the section. He is given the necessary tools and implements, and he is to live in a house adjoining his subsection and to spend the whole of his time on the road work. At stated distances along the road there are to be deposits of road material, approved by the district engineer, and no material except such as is thus deposited can be used on the road. The *caminero* thus has the means at hand to properly look out for the reasonable wear and tear of the road.[57]

While "extraordinary" events—including frequent washouts during monsoon season—would still require more substantial road clearance and rebuilding efforts, the system was built around the individual, geographically distributed responsibilities of the caminero, each tending their own stretch of first-class road:

> It is his duty to keep the vegetation from encroaching; to keep the ditches and culverts clear, so that the water can run off; to keep the surface of the road crowned, so that the water will not lie on the road; to fill all the ruts and depressions in the road with broken rock or other approved material, so as to make drainage perfect and to prevent the beginning of trouble; to keep the road clean and unobstructed, and to protect it against encroachment by neighbors in the way of putting their fences or parts of their houses over it, or the use of the ditches for irrigation purposes or for carabao wallows. He will protect the road from misuse and keep it continually in good condition. [58]

At an estimated cost of 350 pesos per kilometer per year, the system would save three to four times that, Forbes predicted, in preventative maintenance. But while the micro-territorial figure of the *caminero* was pivotal in Forbes's vision of good roads, the provinces would still require the work of road crews and rock crushers, along with technological investments in steam rollers and traction engines, to carry out much of the ongoing surfacing and crowning of roads, and for clearance of debris after storms and landslides. Linking road maintenance to increasingly centralized regimes of inspection under the Bureau of Public Works, the caminero system was also associated with efforts to replace provincial supervisors with district engineers who, Forbes proposed, should make a complete examination of first-class roads in every province every three months, with less frequent inspection required for second- and third-class roads. Establishing a "system of ledgers" for each section of road, detailing "in cold, hard figures" the conditions of roads, bridges, and culverts, and circulating their reports, through central offices, up the chain of authority, ultimately reaching the Governor-General with their recommendations for "any action as may seem necessary." Forbes argued that the road inspection system under the Bureau of Public Works should provide a means of observing not only local practices of road maintenance but also the officials in charge: "The policy should be adopted that any governor, or other provincial official, or any municipal president, or any officer of the Government charged with maintenance of the road, who allows it to deteriorate when he could have maintained it, should be summarily removed from office and somebody else appointed who will more properly perform the duties with which he is charged." Envisioning a world in which each local *presidente* and provincial governor might "make himself a name for good administration by devoting a sufficient part of his energy to the problem of good roads in his district," the caminero system, Forbes hoped, would be both disciplinary and punitive.[59]

Projects of improving, expanding, and maintaining Philippine roads, readily internalized by provincial leaders who stood to benefit from them, were only cost effective and sustainable, Forbes understood, if support was maintained under the double cedula and other road funding provisions (or indeed, increased, since the construction of new roads would continually increase maintenance demands). His efforts to guarantee this funding by making the double cedula "permanent" remained central to an ostensibly apolitical agenda as Forbes prepared to ascend to the Governor-Generalship in November 1909.[60] As Forbes put it in a letter to President Taft prior to his inauguration, the "one essential prerequisite to success in any form of government is material development as providing the sinews of society. I believe it is easier to divert their minds from political matters by giving them something to think of than by entering into arguments with them for the purposes of refuting their statements or by making promises which I am not in a position to fulfill."[61] Despite forestalling questions of independence as fundamentally outside his purview, in nearly four years as the Insular government's chief executive Forbes would find it difficult to stay above the fray, even on inauguration day.

Forbes recorded few observations of the inaugural events in his journal. He mentions that Yu Dong, who would stay on as Forbes's steward and household manager after the move to Malacañang, had laid out his black dress suit for the occasion but that he would instead wear his standard white linen for both daytime and dinner events; and that he managed to squeeze in some polo around the events. But his inaugural address offers a crafted glimpse of the imperial moment from the top-down, setting forth the new governor's policy agenda, political philosophy, and ceaseless sanctimony through a "politics of no politics" and "empire of no empire." Its "keynote," as Forbes would put it in a letter to Secretary of War Dickinson accompanying the published speech, was "material prosperity."[62]

Not surprisingly, the speech is rife with imperialist platitudes.[63] The "purpose of governance," for Forbes, was "not for our satisfaction, nor for the expression of our theoretical views, but for the *happiness*, *peace*, and *prosperity* of the people of the Philippine Islands," and this relational standard was absolute: "in so far as the people do not enjoy these blessings, we have not yet achieved success." Forbes addresses his audience as "Fellow Countrymen," but the nature of that fellowship is complex in the text: sometimes drawing boundaries between Filipinos and Americans, and other times blurring distinctions. Offering, after McKinley and Taft, the principles of "just and effective government" and pledging to treat the people of the Philippines with dignity and respect under the rule of law, Forbes cannot resist, however, throwing shade on his tropical countrymen. For here was "a climate particularly favourable for some classes of products and capable of yielding vast returns to honest and intelligent expenditure of effort, and yet here we have a people bemoaning their poverty and living from day to day without those reserve supplies so necessary where crops are uncertain," while also lacking access to modern medicine and surgery, nutrition, and housing, "in fact, most of those things which modern civilization believes to be necessary for the happiness of a community." The call for fair wages and good working conditions for "the Philippine laborer" matched well, for Forbes, with the promotion of capital as the solution to virtually every social problem. He states his opposition to the admittance of Chinese labor, because he did not see "any need of it. The Filipino can do all the necessary work in the Islands." And yet Forbes could hardly be accused of pandering in his assessment of Filipinos' capacity for manual labor, adding that "The Filipinos are not strong enough to do the work which is required of able-bodied people."[64] Linking the problem of unchecked intestinal parasites with a loss of vitality and the "power to do work," Forbes advocates for technical solutions, including the expansion of artesian wells as a means of avoiding polluted surface waters, broadly associating human welfare with spatial, social, and corporal transformations of the Philippine geo-body.[65]

Forbes turns repeatedly to Hawaii in the inaugural as a model for what "these Islands" could become, given the adoption of modern agricultural and freight handling methods and the development of suitable infrastructure—better roads, railroads, ports, and regular steamship service. To understand how the two archipelagos had developed differently, Forbes might have looked to the history of informal imperialism and commerce in Hawaii, including the devastating effects of venereal disease on native Hawaiians during this time, or the specificity of planter class and local elite relations, but he focuses instead on a single comparative measure: exports per capita.[66] While the Philippines, with a population of about eight million, had produced exports worth $34 million in 1907, the (incorporated) U.S. territory of Hawaii produced $29 million in exports in 1907 among a population under 2,00,000, making the Hawaiian population *thirty-six times* more productive than the Philippines. "No, it is not labor that is wanted here," Forbes reckons, "it is capital." Overlooked or ignored in Forbes's reliance on exports as fundamental measures of progress, and capital influx as the solution to problems of development, was the fact that the vast majority of Filipino peasants relied on subsistence agriculture for their livelihoods, not cash earnings, as well as the centrality of land use practices as structures of social integration.[67]

"Many Filipinos have a tendency to oppose the introduction of capital into these Islands, either from the United States or from foreign countries," Forbes carped, "fearing lest some who it should militate against the realization of their aspirations. In my judgement it will have the opposite effect." Why wait several generations for the development of adequate domestic capital when "we had better attract for our use the accumulation of wealth already made in other countries, sure that the advantages which flow from them will far more than offset any possible disadvantage due to the fact that some of the profits will leave the country or that the owners of capital will endeavor to influence the administration of the Islands or their political status." But capital, Forbes cautions:

> demands a stable Government. Capital is not particularly interested in the color or design of the flag. It wants just and equitable laws, sound and uniform policy on the part of the Government, just and fair treatment in the Courts. The faith of the United States is pledged that all of these benefits shall be permanently assured to the Filipinos. No capitalist need feel alarmed as to the security of his investment provided it has been made in such a way as to fulfill the conditions imposed by law. The United States stands pledged to the establishment and maintenance of a stable Government in the Philippine Islands, not for the sake of such capital as may be invested here only, but for the sake of the welfare of the Philippine people and of the good faith of the United States before the world. The security of foreign capital is merely an incident in the general security of property and other rights to the Filipino, and both are now permanently assured. It would be a good general policy for us to offer every reasonable inducement to capital to come, and with that end in view, to liberalize our land and mining laws and lessen the restrictions which have hitherto tended to discourage investors. My policy will be to hold out the hand of welcome to all people desiring to engage in legitimate enterprise.

With the country thus capitalized, wages might rise to four to six pesos per day for unskilled workers, Forbes suggests, similar to U.S. norms, up from then current rates of 40 to 60 centavos.

If setting out the welcome mat for capital reflected the radically transformative potential of Forbes's liberal imperialism, the speech remained fundamentally conservative of existing power relations. Forbes also endorsed the work—and distributed power—of the Constabulary, specifying that it should be maintained at its present capacities with no reduction in duties. He touted modest gains in the percentage of Filipinos employed in the civil service (which had risen from 49% in 1903 to 62% in 1909) but did not mention less impressive gains in leadership positions. In Forbes's parlance, however, Filipinos did not constitute a modern nation but a people "capable of nationality"; the U.S. was attempting to create the conditions whereby such a "national existence" would become possible.[68] It was not independence, Forbes argued, but ports, harbors, bulkheads, and the expansion of agriculture along navigable rivers that would produce happy and prosperous people in the country: "To the Filipinos, I say, turn your undivided attention to the material development of your country, and rest confident in the good faith of the United States."[69]

Critical among these paths to material development, "and supplementary to all of these, are the roads, and in the present progress of the work in connection with roads, I find the most happy augury for the future success of the Philippine people."[70] Forbes emphasizes again the use of durable materials and—and effect of permanence—which (as Kramer has described) set the purported Americanness of built environments apart from native counterparts in the landscape.[71] But the materiality of the roads, particularly in tropical environments, was not exactly "permanent"; rather, their maintenance required near constant vigilance, including new labor practices and new modes of territoriality and subjectivity:

> The Government buildings and bridges should always be of reinforced concrete; the roads should be built upon strong foundations, with durable surfacing, and guarded from hour to hour by roadmen to see with jealous eyes that no sign of deterioration is allowed even to appear.

Although the punitive elements of road surveillance that Forbes had advocated, wherein local officials could be dismissed by the Governor-General for failing to keep up with road standards, did not gain traction in Philippine law, the power dynamics of Insular inspection persisted in flexible and informal relations of power. As Governor-General, Forbes would engage in a relentless, geographically distributed charm campaign, personally directing inspection tours across the archipelago, often carried out at a breathtaking pace of travel. In these expeditions, to which we turn in the next section, Forbes wielded both the carrot and the stick of colonial rule while constructing, in his journals, a moral topography of "the Islands" that coupled material development, particularly as embodied in road maintenance, with human achievement.

Making a Moral Road Atlas

Regimes of inspection, variously combining elements of the martial inspection, scientific collecting expedition, government census, and goodwill tour, were widespread in Western colonial practice, reflecting and reproducing relations of power and mobility that were both material and symbolic. For Dean Worcester, the longstanding Secretary of the Philippine Interior, the "annual inspection trip" through Mountain Province (among other extensive travels) was understood as a key element of governance, providing an institutionalized setting for face-to-face interactions that enabled a range of personal or charismatic, disciplinary, and transactional political relations with a geographically distributed population. The inspection tours also offered the experience—and performance—of an empire of masculine adventure, projecting an image likely seen as advantageous or even necessary for non-military colonial officials.[72] Worcester regularly lead mounted pack trains with Constabulary officials and escorts across the Cordillera, commonly accompanied by local provincial and sub-provincial governors and sometimes including other Insular government officials and visiting dignitaries, to engage with, in often elaborate performances of tutelage and benevolence, heretofore exotic subject peoples. These localized imperial visits have been well-documented, centering on the production and categorization of thousands of "ethnological" photographs, bodily measurements, and cultural artifacts collected by Worcester and his team, along with his distribution of presents that helped make the visits popular.[73] The visual appearance of neatness and order—and the structural bias of inspection practices toward visual, superficial observation—shaped the knowledges produced, and relationships forged, under such inspection regimes. In turn, these practices informed expectations and preparations in the places visited, as towns and villages with spotless streets and conspicuous displays of welcome, including banners or decorative arches on roads and trails, were known to be appreciated.

With Baguio serving as a convenient point of entry, Forbes sometimes accompanied Worcester on these missions as a Philippine Commissioner and even as Governor-General. In April 1910, just four months after his inauguration, Forbes started an extensive "horseback trip with Worcester through his mountain country," relaying to former Governor-General Luke Wright, Forbes's less mobile predecessor, that he expected "to be gone nearly a month, as besides visiting all the wild tribes in Northern Luzon, I am going to visit the provinces of Cagayan and Isabela, which I have never seen, Ilocos Norte, Ilocos Sur, Abra, and La Union."[74] Across colonial departments, the subjects of Worcester's and Forbes's inspections, including the conditions of roads and trails and general appearance of the landscape (and, moreover, the colonial ethos of inspecting *everything*), overlapped on a range of tours in different parts of the archipelago. For Forbes, the regular inspection of roads and other public works under his purview as Secretary of Commerce and Police, which included visits to jails, schoolhouses, Constabulary stations, hospitals, and other facilities, was expanded to consume, ideally, a third of his time and efforts as Governor-General, as he would later explain to Secretary of War Henry Stimson.[75] Meanwhile, the pomp and politics of local visits—including a wave of ceremonies for the inauguration of new bridges—increased alongside the tours of inspection, featuring children's folk dancing performances and sporting exhibitions, including basketball and baseball games with Forbes himself sometimes participating in the latter. A busy schedule of speeches, banquets, private dinner parties, and fancy balls was regularly articulated with the inspection visits. And while Forbes liked to complain of the endless Filipino courses that had him "fletcherizing at a banquet with six kinds of meat and three of fish," he evidently saw the social events as critical to the "road campaign," and part of an ongoing, geographically-distributed charm offensive through which he hoped to establish a more permanent basis for continued road-building and maintenance programs.[76]

U.S. Coast Guard vessels were favored for coastal and inter-island travel, allowing top-tier bureaucrats like Forbes and Worcester the use of mobile (and comfortable) bases of operations for sometimes lengthy, intricately planned tours of inspection. For trips closer to home, Forbes had, by February 1910, shipped a new Packard touring car from Detroit, and with his chauffer "the faithful Pedro" at the wheel, he plied the roadways of Luzon, inspecting as he travelled.[77] The achievements of material development, and sometimes its shortcomings (always seen as problems of execution not policy) were easily grasped from the road, Forbes believed, on the satisfying terrain of visual observation. On a 1908 inspection of the island of Negros, for example, Forbes started for Bacolod over a "fine road recently fixed, great cement bridges that with proper care will last a century and only just finished in accordance with our new policy."[78] The following year, already serving as Acting Governor-General in the absence of James F. Smith, Forbes inspected new water works of the City of Manila at Maraquina and gushed:

> I love to see the real permanent construction going on. We went by automobile all the way. I could see the results of my road campaign … deposits full of approved road material every few feet, and every little while a caminero busily, but not very skillfully, at work, at least so I thought. Everywhere was evidence of progress—a huge cement building at McKinley as we went through, a fine new bridge over the Pasig at Pasig with road and railway, a stunning new provincial building at Pasig designed by Parsons, a new jail, and then all the water works construction.[79]

In Negros, where Forbes observed a "veritable enthusiasm" for road and bridge building, he inaugurated two new bridges, "great noble structures of reinforced concrete."[80] Later, he read in the landscape a "noble new road" connecting Nueva Vizcaya to Isabela in northern Luzon, an "epoch-making event for both provinces."[81] Indeed, after a 1911 visit to Mindanao, there seemed no limit to the possibilities of road-building, which functioned as both a *producer* and an *indicator* of progress:

> creeping slowly mile by mile, year by year, opening up a world of wealth, waiting only the opportunity that this road brings, to be useful to mankind, order, and happiness, and progress and wealth follow, and poverty, misery, superstition, and danger to life and limb, flee before it. Little by little this road will branch out and connect up till the whole of the wonderful island of Mindanao is netted with roads, and its area, great as Indiana, becomes as orderly and available and many times more populous and productive. If only we can be left alone here and the present policy be allowed to get crystallized before it is taken out of the creative hands that are now bringing it to maturity.[82]

Whatever the truth of Forbes's retentionist claims, his drive to make his colonial subjects see road development the way he saw it, and to crystallize these views in the material landscape of first- and second-class roads, remained a tenuous endeavor. Those with a stake in the matter, who wanted what Forbes wanted, or saw in roadbuilding a means of consolidating their own profits and power, tended to agree.

Still, Forbes was wary of superficial cooperation on the roads policy and incensed when he detected feigned busy-ness on the roads during inspections or merely last-minute preparations, and relished his own capacity, by interviewing local observers, for seeing through such traps of appearance. While punitive dimensions of the road bible were not enshrined in law, under Forbes norms of informal governance emerged in which, among local politicians and colonial officials, active road builders (and maintainers) gained favor. At Siquijor, Forbes praised the "energetic road-building" of the sub-Governor, James Fugate, an American ex-soldier and teacher then climbing the ranks of the colonial cadre.[83] As Forbes would describe him on a subsequent visit to the island in 1911, Fugate was a "quiet little American who is doing wonders with these people and has literally covered the island with good trails and roads, largely by voluntary labor."[84] Conversely, the failure of leaders to adequately maintain the roads was viewed as a serious character flaw, and an explicit rejection of the territorial bargain—and moral economy—of colonial development. At Cagayan province at the northeastern tip of Luzon, for example, Forbes found, "The roads here, as is to be expected with a weak governor, are in the worst condition of any we've seen."[85] Such moral geographies, always racialized, were sometimes read into crude physiognomies, as when Forbes described a "delegation from Tayabas" who visited him "to complain of their local presidente. They charged him with graft, in which they initiated collusion on the part of the provincial governor, a mean, weak-faced looking man named Lopez, whom I excoriated a day or two ago for letting his roads get into disrepair."[86] In contrast, Pampanga was "a good province," wherein Colonel Blanco, "tall, handsome, and a mestizo," an erstwhile Spanish citizen, informed Forbes that "the road over which we traveled was now passable for the first time in forty years at this season, and we crossed three streams over new steel bridges just completed."[87]

Though typically, in his journal entries, Forbes found fault with Filipino officials who had *allowed* roads to fall into disrepair, severe judgements were not limited to Filipinos. Visiting the Pagbilao–Atimonan Road in Bicol, Forbes confessed that he was "disgusted with the whole thing. Another example of outrageous incompetence on the part of the Bureau of Public Works" prior to his appointment. Here, he recalled, "Vogelsang, the engineer, paid for his foolishness with his life, as a slide on the mountainside up which he was putting his road buried him alive, or killed him as it struck him, I don't remember which; but the punishment was by no means too severe for the offense, for the road was in the wrong place, with sharp curves and grades and there are good routes to be found by going after them."[88] Forbes was equally parsimonious in his efforts to discipline Filipino provincial officials who did not meet their road building or maintenance expectations. As he once played hardball with leaders of Ilocos Sur: "In short I told them that if they failed to build and maintain roads, all public works in the province would be discontinued by public order immediately."[89] We can presume that this was not the only time Forbes issued such a threat. Provincial officials were quick to grasp the implicit messages of Forbes's road inspection regime, or at least to tell him what he wanted to hear. Recounting a conversation at a party hosted by Governor Tinio of Nueva Ecija, Forbes noted that Tinio said he had "wished that I had become Governor before, because from what he could see of the road work it was the life of the people. He said he'd been watching how it worked in the provinces."[90]

With the support of his "strategy board"—constituted by "the Americans with portfolios on the Commission"—Forbes called for an extra session of the Legislature, and with Osmeña's help, was at last able to garner adequate support for the permanent double cedula tax provision, achieving with its passage, he believed, an "assured, permanent road development."[91] While ongoing appropriations for roadwork had depended on the provincial boards signing the double cedula into law, Forbes believed that the measure also gave him "good reason for refusing to approve any resolution repealing the law in any province that our road campaign is clinched."[92] It was, Forbes reflected triumphantly, "The greatest of all accomplishments. We have made the double cedula active continuously from year to year … This is the culmination of my five years' work on roads. The money is now permanently available; it can't be stopped."[93] Concluding on a maudlin note, he added: "If I should die now, I think I could justify my existence as I think this one five-years drive at roads, riveted and double-clinched by this law, makes a work that is worth the whole sacrifice, the whole strain, and the whole, really, of a life."[94] The perceived permanence of the provision, however, by no means signaled the end of his inspection visits. Following one January 1911 inspection, for example, Forbes noted that while the road remained "in fine shape" he did not like to see the camineros using loam for surfacing instead of "deposits of road materials approved by engineers at various intervals."[95] The result, Forbes later recalled, would have been the kind of harangue with Public Works or local officials that he was known for: "I stood first, last, and all the time for scientific road maintenance. This sort of inspection continually going on was the kind of thing that kept the Islands up to good administration."[96]

By September 1911, Forbes could describe, in a letter to Secretary of War Stimson, an extended "general inspection" model geared toward a "thorough overhauling to the administration of the provinces on the spot."[97] The inspection tour was ambitious, including in Forbes's entourage members of the Philippine Commission and bureau chiefs who worked in the provinces or their assistants. Forbes commandeered the cable ship *Rizal* as the inspection tour's base of operations, thus "cutting out" what Forbes now viewed as excessive "entertainments and receptions."[98] Thus, turning the geography of territorially defined representative democracy "inside-out," in a sense, through traveling Insular government inspections, Forbes attempted to innovate new ways for the Insular government to travel *to* the people. Though it is not clear that any additional, formal powers were bestowed on the general inspection, Forbes continued to exercise proconsular powers on the tours, including visits to local jails, in which Forbes often met with prisoners about their cases and experiences in the justice system. Forbes sometimes offered clemency to a host of prisoners awaiting trial while denying it to others.[99] While the "general inspection" functioned, in some ways, as an imperial fantasy of power, the social relations it engendered or reproduced were nonetheless real, though the dynamics of power were largely informal and transactional, as provincial officials and other elite actors, while telling Forbes what he wanted to hear, also tried to make *him* want what they wanted.[100]

"It is astounding what a hold my policy of road construction has taken on the people," Forbes insisted to Stimson, noting that petitions for additional roads and structures were often presented to him during the visits.[101] The next month, after viewing "excellent roads and lovely country" on the way to Nueva Caceres—an indicator of the rectitude of governing officials—Forbes was delighted to encounter popular enthusiasm for road-building in:

> a public session in which the orators declared for roads and roads and roads, insisting with oratorical vehemence approaching passion on their absolute necessity for the life of the people. This province has more roads per capita than any other and has done better in the matter

of roads than almost any other, which shows that the road movement is sure to gain headway by reason of its own success, as the more roads people have the more they want…[102]

Such support was not universal. Responding, during a 1910 visit by Secretary of War Dickinson, to "a conference held yesterday with the 'boys' of the press," following his inspection of Philippine roads with Forbes, in which Dickinson had insisted that "the main thing for the country is good roads, which facilitate commercial exchange and its progress," the newspaper *El Ideal* objected to the priority that roads had been afforded—by Dickinson and in the Insular budget. Given conditions of "economic crisis" in the archipelago, the anonymous editorial asked, why were roads the country's "main necessity"? While the article allowed that the Governor and Secretary of War were acting in good faith, it questioned the belief "that in the Philippines the opening of a good path or the laying of some iron line can produce the same miracles as in the U.S., where capital, always alert, brings prosperity to open points of traffic." For if the roads policy was a safe investment for "rich villages," the outcomes of road building in advance of material prosperity, "for a people, poor like ours," were more dubious, indicating "an error of principle; there's a factor-in-the-box. What should have been the last has come first."[103] But the matter of who stood to benefit from Forbes's "road movement" was, like a host of other matters of colonial policy, a socially and geographically uneven question.

The problem, as Forbes saw it, was not road policy but "native newspapers" and their relentless vilification of the Insular government that constituted "the greatest menace to the success of the administration here."[104] It pained him that the "noble work" on the roads was not more widely celebrated, adding to a pervasive sense of besiegement, reflected throughout Forbes's public and private texts, and the notion that bad Filipino actors, if not the Filipino people as a whole, were to blame for impeding American-style economic progress in the archipelago. Perhaps he had even come to agree with his friend Charles Eliot who, in an exchange following Forbes's inauguration, had gently pushed back on Forbes's enthusiasm having attained the Governor-Generalship:

I congratulate you on having 'got to a place where I can do things'. That is a fortunate state, provided that the people for whom you are working want to have your things done, and believe that they are themselves to be helped thereby. One of the most difficult and thankless forms of labor in this world is trying to do good to people who resent and obstruct your efforts. It is almost impossible to confer even the plainest benefits on unwilling people.[105]

Just so, even as Forbes celebrates the American colonial project in the Philippines, there is a specter that haunts his journals and letter books—and the "moral road atlas" that emerges within them: the specter of an empire *ruined* by its openness to democracy, occurring even at its moment of triumph. And yet what Eliot (like Forbes) does not consider or takes as self-evident are the assumptions, prejudices, and values built into the work of "doing good," such as the construction of roads laid out to subsidize an extractive export economy designed to benefit large landholders, or to carry American colonials and their allies to their summer retreat. Forbes is compelling, and was in some ways effective, as an *agent* of liberal empire because his ambitions, for his own career and for "material development" in the Philippines, were compatible with those interests, and he was equally full of contradictions. Like the American empire of no empire in the Philippines that he helped to reproduce, Forbes continually sought to convince others—and himself—that he *was* good.

Notes

1 William Cameron Forbes, "Journal" (entry 5/30/1908), First Series, Vol. III, p. 30, W. Cameron Forbes Papers (1930), MS Am 1365, Houghton Library, Harvard University.
2 Forbes, "Journal" (entry 5/30/1908), Vol. III, p. 39.
3 Forbes, "Letter of the Secretary of Commerce and Police to All Provincial, Municipal, and Other Officials Relative to the Present Road Policy in the Philippine Islands," June 16, 1908, Department of Commerce and Police, Manila. W. Cameron Forbes Papers, bMS Am 1364.2, Box 2. The file retains a translated printing of the letter in Spanish but none in Philippine languages or dialects.
4 See, for example, the discussion in Taft, "Special Report of WM. H. Taft Secretary of War. To the President on the Philippines" January 23, 1908, War Department. Washington, DC: Government Printing Office, p. 64.
5 Forbes, "Letter of the Secretary of Commerce and Police."
6 Ibid.
7 Ibid.
8 Henri Lefebvre, *The Production of Space*, trans. Donald Nicholson-Smith (Oxford: Blackwell, 1991), p. 245.
9 Lefebvre, *Production of Space*, p. 44.
10 "First class roads" were defined as well-graded and surfaced with approved materials, "thoroughly drained and constantly maintained." Expectations were that bridges and culverts should be permanent structures, with ferries in place capable of transporting automobiles if not, and that the roads were "continuously passable at all times with possible exceptions during typhoon periods." No. 6 Road Map of the Philippine Islands, Compiled from Reports of the Engineers of the Bureau of Public Works and Other Official Sources, June 30, 1919, Insular Government, Manila. RG 330/21/18/5-3, Cartographic and Architectural Section, National Archives and Records Administration (NARA), College Park, Maryland.
11 Taft, "Address by Wm. H. Taft, Secretary of War, at the Inauguration of the Philippine Assembly, October 16, 1907" January 23, 1908, War Department. Washington, DC: Government Printing Office, p. 81.
12 Forbes, "Journal" (entry 5/30/1908), Vol. III, p. 30. Of the two million pesos (worth about one million 1908 dollars), three-quarters was slated for roads and one-quarter for irrigation purposes. See also Forbes, "Journal" (entry 5/26/1908), Vol. III, p. 28.

13 Forbes, "Letter of the Secretary of Commerce and Police."
14 As Michael Cullinane notes, Forbes saw in Osmeña's election to the speakership the ascendancy of a politician who could be controlled. Cullinane, *Ilustrado Politics: Filipino Elite Responses to American Rule, 1898–1908* (Quezon City: Ateneo de Manila University Press, 2003). On the continuing political fortunes of the Osmeña family in Cebu, see Resil B. Mojares, "The Dream Goes On and On: Three Generations of the Osmeñas, 1906–1990" in Alfred W. McCoy (ed.), *An Anarchy of Families: State and Family in the Philippines* (Madison, WI: Center for Southeast Asian Studies in cooperation with Ateneo de Manila University Press, 1993), pp. 311–346.
15 Greg Bankoff, "'These Brothers of Ours': Poblete's *Obreros* and the Road to Baguio 1903–1905," *Journal of Social History* 38 (2005): 1047–1072.
16 In harder to reach mountainous villages and rancherias, peasants may have reengaged earlier responses to the Spanish *polo*, disappearing into the bush when government officials appeared to collect taxes or labor.
17 Justin F. Jackson, "'A Military Necessity Which Must be Pressed': The U.S. Army and Forced Road Labor in the Early American Colonial Philippines," in M.M. van der Linden and M. Rodríguez García (eds.), *On Coerced Labor* (Leiden: Brill, 2016), pp. 127–158.
18 Greg Bankoff, "Wants, Wages, and Workers: Laboring in the American Philippines, 1899–1908," *Pacific Historical Review* 74 (1998): 59–86; Bankoff, "These Brothers of Ours"; Jackson, "A Military Necessity."
19 Forbes, "Journal," ("Wood-Forbes Mission: Interviews, selected and commented on by W. Cameron Forbes," no date), Second Series, Vol. II, p. 220.
20 Forbes, "Journal," Second Series, Vol. II, pp. 220–221.
21 Philippine Commission, "Report of an interview between Commissioners Worcester, Wright, Smith, and the President [Taft] sitting in committee by authority of resolution of The Philippine Commission of June 1, 1903, and Mr. N.M. Holmes, engineer in charge of work on the Benguet Road, Baguio, Benguet, June 2, 1903," RG 350/Box 274/File 2373, NARA.
22 Philippine Commission, "Report of an interview."
23 Ibid. Holmes at times paid Filipino road workers five cents and a 2.25 pound sack of rice per nine-hour workday (as noted in Jackson, "A Military Necessity.")
24 Philippine Commission, "Report of an interview."
25 Ibid. Five-hundred Chinese workers, who could be furnished for one peso per day by a boss who disciplined the labor himself, were worth 1,000 Filipinos, the frustrated Holmes insisted.
26 Ibid.
27 Kennon previously served as military governor of the province of Ilocos Norte in 1901 and oversaw construction of the Iligan to Lake Lanao road in Mindanao from 1901 to 1903. Today the Marcos Highway competes with Kennon Road for the best route between Manila and Baguio City, though from personal experience, in a small sample size, I can confirm fewer landslide delays on the Marcos.
28 Bankoff, "These Brothers of Ours."
29 Ilocanos, by October 1903, constituted some 2,000 of 2,800 Filipino workers employed on the road. Bankoff, "These Brothers of Ours." While conditions for workers generally improved under Kennon, including amenities for workers, living in large camps of up to 1,000, such as weekly dances, saloons, and cockpits for recreation, along with improved medical facilities, labor strife (along with the dangers of rock work) persisted.
30 The success of the road, coming in at a cost more than 30 *times* the original estimate and subject to frequent washouts and road closures, would remain an open question. Robert R. Reed, *City of Pines: The Origins of Baguio as a*

Colonial Hill Station and Regional Capital (Berkeley, CA: Center for South and Southeast Asia Studies, University of California, 1976).

31 Forbes, "Journal" (footnote to entry 9/17/1904), First Series, Vol. I, p. 70. The reassessment of Kennon was amended as a footnote in the journal prior to the 1930 donation of the Forbes Papers to the Houghton Library.

32 Forbes, "Journal" (entry 9/17/1904), p. 68. Forbes also expressed satisfaction with Kennon's work in the pages of *La Democracia* shortly after his first visit to the road. "Hablando con Mr. Cameron Forbes: La cuestion de las carreteras" *La Democracia* September 22, 1904. W. Cameron Forbes Papers (1930), b MS Am 1364.4, Box 4.

33 Forbes, "Journal" (footnote to entry 9/17/1904), p. 70. Lyman Kennon returned to North America and continued his military career after his Philippine posting. As Forbes offers in epilogue, Kennon was "made a major-general in the Great War and given a brigade to train. He was not, however, deemed fit, whether physically or otherwise I do not know, to go abroad and was relieved on the eve of sailing … Kennon went back to the hotel and died, some say of a broken heart, and others say by his own hand." The *New York Times*, conversely, notes Kennon's cause of death only as a "brief illness," and his rank as Colonel. "Col. L.W.V. Kennon Dead" *New York Times* September 10, 1918, p. 9.

34 Forbes, "Journal" (entry 9/17/1904), p. 68.

35 Forbes, "Journal" (entry 9/17/1904), p. 71.

36 According to Kennon (in Bankoff, "These Brothers of Ours"), 46 nationalities contributed to the road's construction, reflecting a surprisingly cosmopolitan labor force at the onset of the American imperial moment.

37 Forbes, "Journal" (entry 9/17/1904), pp. 70–71.

38 Forbes, "Journal" (entry 9/17/1904), p. 68; Forbes "Journal" (entry 9/05/1904), Vol. I, p. 61.

39 Forbes, "Journal," Vol. 1, pp. 61–201. The road's "completion" was itself achieved symbolically on January 30, 1905, when Kennon himself drove an oxcart over the last, still rough stretch of road to Baguio in time for workers to cash in on a friendly wager of cigars, beer, and brandy made by Kennon and Forbes, a motivational technique the latter attributed to Grandfather Forbes and his Midwestern railroad stories. Forbes, "Journal," (entry 2/07/1905), Vol. 1, p. 145.

40 Forbes, "Journal" (entry 9/17/1904), p. 69.

41 Alex Lichtenstein, "Good Roads and Chain Gangs in the Progressive South: 'The Negro Convict Is a Slave'," *Journal of Southern History* 59 (1993): 85–110; see also Nicholas Gerstner, Adrienne R. Hall, and Scott Kirsch, "An Archive of Good Roads and Racial Capitalism in North Carolina," *Human Geography* (forthcoming).

42 Forbes, "Journal" (entry 12/20/1904), p. 125. Forbes also incorporated prison labor into proposed industrial and technological solutions to the challenges of road maintenance in tropical environments, such as the production of wide cart tires, which would do less damage than narrow tires to muddy roads and could be produced at Bilibid Prison for about half the cost.

43 "Permanent Road, Bridge, and Building Construction in the Philippines" *The Far-Eastern Review* August, 1907, p. 82.

44 Jackson, "A Military Necessity," p. 132.

45 As Jackson notes, American officials, like Constabulary Lt. Jefferson Davis Gallman, governor of the Ifugao sub-province (and nephew of the namesake Confederate president), boasted of their ability to conscript at least 20,000 laborers each year for stints of road labor and other civil works projects. Jackson, "A Military Necessity," p. 157.

46 Eliot wrote to Forbes, following the latter's January 12, 1906 talk at the Tavern Club in Boston, that he found the lecture "clear and interesting" but took exception to the corvée. "Forced labor must be either very ineffective and

demoralizing or, if effective, it is compelled by some sort of public cruelty, and is therefore extremely demoralizing. All the other American performances in the Philippines which you described were civilizing; this one is sure to be degrading." Eliot to Forbes, January 13, 1906. W. Cameron Forbes Papers, bMS Am 1364, Box 2, Houghton Library, Harvard University.

47 Forbes, "Letter of the Secretary of Commerce and Police."

48 Ibid.

49 "Road Building in the Philippines" *The Far-Eastern Review* vol. IV (1908), p. 308. Viewpoints on transportation infrastructure in the *Far-Eastern Review*, an English language commercial and engineering trade journal founded by American George Bronson Rea in 1904 with support from the War Department, largely mirrored Forbes's own. See also "Public Works in the Philippine Islands" *The Far-Eastern Review*, Vol. V (1908), p. 237.

50 To borrow from Paul A. Kramer's description of an earlier moment of collaboration in *The Blood of Government: Race, Empire, the United States, & the Philippines* (Chapel Hill, NC: University of North Carolina Press, 2006), p. 171.

51 Far-Eastern Review, "Permanent Road, Bridge, and Building Construction," p. 82.

52 The provincial boards consisted of a provincial governor, elected by the municipal councils (themselves elected through narrowly circumscribed electoral processes), treasurer, and supervisor of public works; the latter positions were, until 1907, both appointed by the Governor-General. After 1907, municipal councils gained authority to appoint a second member of the provincial triumvirate. As Kramer describes it, "this structure … rooted the new American colonial state in entrenched rural power structures, surrendering local colonial politics to the *principales*, guaranteeing them control over municipal councils and provincial governorships and satisfying their desire for control of their constituents and labor forces. At the same time, this structure allowed American appointees to retain fiscal authority." Kramer, *Blood of Government*, p. 173.

53 Far-Eastern Review, "Permanent Road, Bridge, and Building Construction," p. 82.

54 Taft to Forbes, November 30, 1907. W. Cameron Forbes Papers, bMS Am 1364, Box 7, File 291. Houghton Library, Harvard University.

55 Forbes, "Letter of the Secretary of Commerce and Police."

56 Ibid.

57 Ibid.

58 Ibid.

59 Ibid.

60 Forbes served as Acting Governor-General (in James Smith's absence) for about six months prior to his appointment.

61 Forbes to Taft, October 12, 1909. W. Cameron Forbes Papers, MS Am 1366.1, Confidential Letter Book No. 1, Houghton Library, Harvard University.

62 Forbes to Dickinson, November 27, 1909. W. Cameron Forbes Papers, MS Am 1366.1, Confidential Letter Book No. 1, Houghton Library, Harvard University.

63 All quotations in this section, unless indicated, are from the "Inaugural Address of the Honorable William Cameron Forbes November 24, 1909." Manila: Bureau of Printing, 1909, W. Cameron Forbes Papers, bMS Am 1364.4, Box 2. Houghton Library, Harvard University.

64 "Inaugural Address." The Filipino body was a common theme for Forbes. Developing "the physique of the people, so that it is physically possible for them to do an able-bodied man's labor," he had lectured the 1908 Lake Mohonk conference in New York, was the first element of a three-part solution to the question "What had best be done for the material advancement of the Philippines," followed by improved means of communication and exchange

and stimulation of modern lines of production, which would together make Philippine labor more efficient. Forbes, "Memorandum for a Speech at Lake Mohonk Conference," October 22, 1908. W. Cameron Forbes Papers, bMS Am 1364.4, Box 3, Houghton Library, Harvard University. By contrast, Forbes praised the Igorots of Mountain Province as "tremendous workers, some of them of beautiful physique, but isolated from the world." Forbes to Dickinson, January 8, 1910. W. Cameron Forbes Papers, MS Am 1366.1, Confidential Letter Book No. 1, Houghton Library, Harvard University.

65 "Inaugural Address."

66 On the impacts of European diseases on Hawaii and other Pacific islands, see, David Igler, *The Great Ocean: Pacific Worlds from Captain Cook to the Gold Rush* (New York: Oxford University Press, 2013), pp. 43–72.

67 Rodney J. Sullivan makes a related point, focusing on Worcester and U.S. colonial views of Philippine agriculture, in a slightly earlier context, in Sullivan, *Exemplar of Americanism: The Philippine Career of Dean C. Worcester*, Michigan Papers on South and Southeast Asia 36 (Ann Arbor, MI: Center for South and Southeast Asian Studies, University of Michigan, 1991), p. 101.

68 "Inaugural address," p. 18.

69 "The United States is strong, determined, fixed in her policy," Forbes added, "and not to be dissuaded or coerced."

70 Building good roads was thus a "prime necessity," as Forbes elaborated as part of a media campaign that followed the inauguration, that would stimulate agricultural production in the Philippines by removing "the obstacles which prevent the rapid development of the people." Governor-General Forbes, "Good Roads, A Prime Necessity" *Philippine Resources* 1: 2 (1909). Unpaginated. Philippine National Library.

71 Kramer, *Blood of Government*, pp. 309–310.

72 On forms of "bureaucratic masculinity" among the American colonial cadre in the Philippines, see Karen R. Miller, "'Thin, Wistful, and White': James Fugate and Colonial Bureaucratic Masculinity in the Philippines, 1900–1938," *American Quarterly* 71(2019): 921–944.

73 Sullivan, *Exemplar of Americanism*, pp. 141–164; Mark Rice, *Dean Worcester's Fantasy Islands: Photography, Film, and the Colonial Philippines* (Ann Arbor, MI: University of Michigan Press), pp. 1–79.

74 Forbes to Wright, April 20, 1910. W. Cameron Forbes Papers, MS Am 1366.1, Confidential Letter Book No. 1, Houghton Library, Harvard University.

75 Forbes to Stimson, May 31, 1911. W. Cameron Forbes Papers, MS Am 1366.1, Confidential Letter Book No. 1, Houghton Library, Harvard University. Although this figure is perhaps an exaggeration intended to impress the new Secretary of War, the geographical extensiveness of his travels, and at times the relentlessness of his work habits, are evident in Forbes's journals.

76 Forbes, "Journal" (entry 7/02/1909), Vol. III, p. 186.

77 Forbes, "Journal," (entry 2/18/1910) Vol. III, p. 419; (entry 6/28/1913) Vol. III, p. 275.

78 Forbes, "Journal," (entry 4/17/1908) Vol. III, p. 2.

79 Forbes, "Journal," (entry 6/21/1909) Vol. III, p. 175.

80 Forbes, "Journal," (entry 7/26/1909) Vol. III, p. 257.

81 Forbes, "Journal," (entry 4/22/1911) Vol. IV, p. 329.

82 Forbes, "Journal," (entry 8/29/1911) Vol. V, p. 19.

83 Forbes, "Journal," (entry 7/15/1909) Vol. III, p. 219.

84 Forbes, "Journal," (entry 9/04/1911) Vol. V, p. 28. Within a few years of the visit, Fugate had been accused of seduction of minors, licentiousness, and perversity, and he turned to missionary work in Mindanao, returning to colonial administration before he was beheaded in Cotabato in 1938, killed in a manner that suggested the settlement of grievances. Miller, "'Thin, Wistful, and White.'"

85 To make matters worse, baseball and basketball events had been called off due to rain and "the dinner was a horrible affair served cold in the provincial building, and we were quite glad to get back to the ship after a rather pleasant ball." Forbes, "Journal," (entry 8/31/1911) Vol. V, p. 23.
86 Forbes, "Journal," (entry 7/26/1909) Vol. III, p. 236.
87 Forbes, "Journal," (entry 5/21/1910) Vol. IV, p. 144.
88 Whatever Vogelsang's fate, "the scenery was absolutely marvelous," Forbes observed, "and the glimpse of the China Sea form the top of the ridge through a bower of tropical verdure, is a thing worth coming all the way to see. At one spot was a little table spread with whisky and tansan to refresh our throats." Forbes, "Journal," (entry 9/24/1909) Vol. III, p. 306.
89 Forbes, "Journal," (entry 5/20/1910) Vol. IV, p. 159.
90 Forbes, "Journal," (entry 9/14/1909) Vol. III, p. 280.
91 Forbes, "Journal," (entry 2/22/1910) Vol. III, p. 425.
92 Forbes to Smith, April 25, 1910. W. Cameron Forbes Papers, MS Am 1366.1, Confidential Letter Book No. 1, Houghton Library, Harvard University.
93 Forbes, "Journal," (entry 4/21/1910) Vol. IV, p. 45.
94 Ibid.
95 Forbes, "Journal," (entry 1/05/1911) Vol. IV, p. 269.
96 Ibid.
97 Forbes to Stimson, September 15, 1911. W. Cameron Forbes Papers, MS Am 1366.1, Confidential Letter Book No. 1, Houghton Library, Harvard University.
98 Ibid.
99 Visiting Nueva Caceres in the Bicol region in October, 1911, Forbes "turned loose all the women in the jail but one just convicted who had one sick child and was expecting another child," for whom he "went to the judge to see if he saw any necessity of her service sentence." On the same visit, "There was a long line of men prisoners, every third one of whom was in for gambling. Almost all wanted pardons and it took me a half hour to hear all their cases and refuse their requests for pardon." Forbes, "Journal," (entry 10/03/1911) Vol. V, p. 50.
100 Prior to Stimson's appointment, Forbes's penchant for the imperial had led to a kerfuffle, resulting from an exchange with Secretary of War Jacob Dickinson in which Forbes suggested that it was appropriate for the Governor-General to "stretch the law," leading, embarrassingly for Forbes, to Taft's mediation. As Forbes would clarify for Dickinson, "In regards to my words 'stretch the law,' my position is this: that where there are implied powers I am willing to use them, even although the law does not expressly provide that these powers lie, i.e., where the action seems to be necessary for the public good." Forbes to Dickinson, April 5, 1910. W. Cameron Forbes Papers, MS Am 1366.1, Confidential Letter Book No. 1, Houghton Library, Harvard University.
101 Forbes to Stimson, September 15, 1911.
102 Forbes, "Journal" (entry 10/03/1911), Vol. V, p. 50.
103 "Agricultura antes que caminos" *El Ideal* August 11, 1910 (translation mine) in W. Cameron Forbes, "Clippings from the Filipino, Spanish and American Press of the Islands, collected for Hon. Jacob M. Dickinson … on the occasion of his visit to the Philippines, during the months of July, August and September, 1910." W. Cameron Forbes Papers, f MS Am 1365.10, Houghton Library, Harvard University.
104 Forbes to Stimson, August 1, 1911. W. Cameron Forbes Papers, MS Am 1366.1, Confidential Letter Book No. 1, Houghton Library, Harvard University.
105 Eliot to Forbes, March 24, 1910. W. Cameron Forbes Papers, bMS Am 1364, Box 2, Houghton Library, Harvard University.

5 Coda

Insular Empire

The Long Road Home

If in-person inspections, on-the-spot adjudications, fancy balls, and—to the extent they could not be avoided—provincial imperial banquets, had all provided Cameron Forbes with the bona fides of on-site experience and gubernatorial authority, offering the practical political advantages of face-to-face interaction outside Manila, Forbes's long workdays, relentless travel, and rigorous polo schedule all seemingly took a toll on his body. Forbes suffered a number of ailments during his Philippine career, including bouts of dysentery and, as one physician later attested, a "protracted illness stemming from pyonephritis of bacterial origin," a urinary tract infection characterized by inflammation and pus in the kidney, believed to have resulted from a polo injury sustained in Manila in 1907. It likely became a chronic or recurring condition; we know that Forbes experienced a similar internal infection in the Fall of 1911.[1] Writing to General Clarence Edwards, the longtime chief of the Bureau of Insular Affairs, in November 1911, Forbes described a "slow and tortuous road to recover" on which he "did not know how sick I was … although the fever and pains in the head were very serious … unbeknownst to me … one of my kidneys was pretty seriously out of whack and might have put me out of business personally."[2] Relieved that he would not require an operation, Forbes was now "recovering swiftly," but would be kept from polo for a few more weeks. Three weeks later, he remained in poor health, relating doctor's orders to stay in bed, on one occasion, and on another, the inspection of severe storm damage around the Benguet Road, confessing that he "submitted to a chair borne by four Igorots, who took us by the scene of the landslide which recked the road."[3]

The state of the devastated road and bridges did little to cheer him. Forbes "mourned greatly over the sad condition of these splendid roads, which were much damaged by the terrible floods which had swept over them."[4] Surveying the wreckage, he observed that a "whole mountain indeed came down and the scene is vast and terrible in its aspect as well as in its consequences. Our two longest and biggest steel bridges are gone; one of them lies exposed in the gravel, a terribly bent, twisted and rolled mass of iron

DOI: 10.4324/9780429344350-6

work, inside of which are wedged pieces of trees and rocks."[5] Forbes now recognized (though surely he had witnessed it for years) the vulnerability of the road that the storm had laid bare; the Benguet Road, as he pronounced in a letter to Secretary of War Henry Stimson, was "not safe; it will certainly be closed during high water and undoubtedly impassable for several months each year."[6] Confiding in Stimson his concerns about Taft's electoral vulnerability, Forbes worried that a future, Democratic-appointed Insular Government might "adopt the very foolish expedient … of abandoning Baguio in order to get a little cheap applause just at the time when it ought to be clinched."[7] Forbes thus hoped to persuade Stimson to throw his support behind the latest effort to extend a railroad line to Baguio from Dagupan, which would render the investment in the summer capital as "permanent" as the road campaign, adding that Baguio remained an "urgent necessity" for the Filipino people: "They are anemic."[8] Even as road crews struggled to clear debris from where the storm had demolished—and rendered impermanent—one of the Insular Government's most spectacular achievements, Forbes leaned into efforts to concretize the American imprint in the Philippine landscape. And even as his own health showed signs of severe strain, Forbes's zeal for material development, in a world wherein American colonial spaces could simultaneously be made and unmade, persisted as a kind of immortality wish, reflecting anxieties that the American presence in the Philippines, and his own best efforts, might simply be washed away.

As Forbes prepared for his first stateside leave since 1908, his journal entries are infrequent, but after departing Manila on the transport *Sherman* on 18 March 1912, he initiates a series of reflections that the journey and requisite leave from active duties of the Governor-Generalship would allow. Forbes arrived in Nagasaki on March 23 before embarking westward, on the *Chikusen Maru*, across the East China Sea to Shanghai, then north across the Yellow Sea to Port Arthur, on Manchuria's Liaodong Peninsula.[9] Perhaps inspired by Taft's 1908 itinerary, Forbes was to travel westward across Asia on the Trans-Siberian Railway, rather than the customary Pacific route via Yokohama and Honolulu, with plans to visit Moscow, St. Petersburg, Berlin, Paris, and London, meeting friends along the way, before the Atlantic crossing. The long journey, Forbes hoped, would provide much needed time to rest and reflect on "eight years on the Philippine Commission; and two years, ten months, and one week, Acting Governor-General and Governor-General. This latter period continuously and without let up, nor for one minute outside the Islands."[10] But things would get worse before they got better. After the American consul at Nagasaki had given "a dinner that night, with many courses and many wines, and I, fooled by five days' absolute rest and health and apparent strength, ate the full dinner, and the next day an elaborate lunch … when we set to sea I found my mistake, and after a night, spent my day on my back with one slice of toast for nourishment."[11] Still, the opportunity to take stock that the journey afforded was one that he relished.

While acknowledging "how far short of perfection we still are," Forbes was, not uncharacteristically, "very well satisfied with the amount accomplished."[12] Atop the list were the distinctly material transformations of space—roads "built, being built, and maintained"; use of durable materials in public buildings; harbor improvements at Manila, Cebu, and Iloilo; and new railroad construction—which had left deep impressions in the archipelagic landscape.[13] This lasting peace dividend, Forbes insisted, was also reflected in the new public order and better relations between Filipinos and Americans of all classes that had emerged under Insular rule. The conditions for business, trade, and revenue production, along with the expansion of post offices, steamship travel, and cable lines, could be celebrated, while—notwithstanding his plea to Stimson for the railroad guarantee—Forbes confided to his journal that the "success of Baguio seems assured."[14] Lagging land registration and irrigation measures, an inefficient freight system, a dearth of factories processing Philippine export products, and, lastly, the persistence of poverty and disease, were identified among the areas falling short of perfection: "Oh yes!" he concluded the entry in a jaunty key, perhaps talking himself into a return to Manila after the leave, "There's lots to be done."[15]

At Port Arthur, Forbes was well enough, he thought, to amble up a 500-foot hill overlooking the harbor to gaze across the site of a decisive 1904 naval battle in the Russo-Japanese War, but the strain proved too much. As he would later annotate the entry, "this was very likely the blow that killed father." But while Forbes would complain in retrospect that "doctors should have told me my heart was weak," it was only later, seemingly, that physicians had determined that Forbes's infirmity stemmed from "a deterioration of the muscles of the heart, due … to the infection of last autumn."[16] Reaching the railway at Harbin, Manchuria on March 30, Forbes convalesced in a Pullman car, changing over to Russian trains at the border. The next ten days—in this last decade of Romanov rule—were spent hurtling across the Siberian plain in comfort on the famed Siberian railroad.

During this time, his geopolitical imagination fired by the transcontinental travel, Forbes fantasized a narrative of continuing westward progress of (Western) Europeans in the northern latitudes, offering a deeply racialized—and flat-out bizarre—mapping of the imperial geo-body. Russian absolutism, Forbes conjectured harshly, would not be:

> washed out … except in blood. Westward the course of empire takes its way, and when at last the United States fills up and overflows, there will be, I suppose, the sister republic, the United States of Siberia, the same race, language, and interests. With England, this would make a belt right round the world, at the dominant latitude as though a hand were placed over the top of the world.[17]

Arriving in Moscow on April 10, Forbes was impressed by the elaborate military and civil service uniforms, disapproving of the relations between church and state, and appalled by the plumbing. Although, writing in the journal, he feigns appreciation, in Moscow, for a solitude that was unknown to him in the Philippines, in a later annotation Forbes admits that he "didn't feel a bit well" at the time, an ailment that a potent whisky toddy and early trip to bed did not cure: "It was curious but the minute my friends left me alone in Moscow I became the victim of a very deep, black depression, very unusual in me, and similar to those into which I imagine Grandfather Forbes used to fall when hardly anything could arouse interest or spark of his usual animation. I have no doubt it was incidental to my illness."[18] He soldiered on to St. Petersburg (and "the luxury of an absolutely well-appointed bath room") and enjoyed the "cleanliness, loveliness, and perfection of Berlin," but had little appetite for site seeing.[19] By the time he reached Paris, Forbes complained to his journal that he should have "taken an ocean voyage to sleep" rather than the incessant touring.[20]

The salvation of the journey was an invitation to Rudyard Kipling's 350-acre Sussex estate, where Forbes was immediately taken under the care of the pharmacological pioneer Sir Lauder Brunton. "I am very much under all the time," Forbes noted, "with my infirmities and periods of violent coughing, and feared I should wake the whole household, but no one seemed to have heard." Over several days rest and recovery, Forbes savored intimate conversations with Kipling, who read Forbes a new story—later published as "The Benefactors"— set amid a coal strike in Hades that included a pope among its characters. The diplomatic Forbes had questioned whether this might cause "injury to be done to the feelings of our Roman Catholic brethren in seeing a pope portrayed in hell," but observed that Kipling was "in a position not to care."[21] The two shared their closely related views on questions of labor, the benefits of playing solitaire before bed, and, presumably, the "white man's burden," before Forbes boarded the *Lusitania* for New York, arriving, the returning Procurator, on 3 May 1912. Greeted by awaiting family at the pier, Forbes was whisked off to Massachusetts for a few days homecoming before boarding a night train to Washington for meetings with Stimson and, eventually, Taft.[22] On May 14, Forbes travelled to New York to attend a banquet in his honor at Sherry's on Fifth Avenue, where he would encounter the parade of elegant courses and imperialist musings with which this book began.

Compared to the soaring, planetary rhetoric of that evening, it had been a long slog across the earth's surface for Forbes, albeit a comfortable one. For readers of *American Colonial Spaces in the Philippines*, it has also been a long road back to Sherry's in time and space, and not always a linear path. Looking at key forms of sovereignty (territory), knowledge (maps), aesthetics (landscape), and circulation (roads) "through" the production of space, the book has, in the intervening chapters, traced the efforts of an American colonial regime to make space for empire in the Philippines during the "long first decade" of U.S. rule. But creating a space, or variegated spaces, for empire's survival was always a complex and unstable achievement, not a monolithic exercise or one-way projection of power. American colonial spaces reflected and reinforced power inequalities but like other state spaces, they withered away at "weak points."[23] They engendered new contradictions: an empire 'ruined' by democracy, and reshaped by oligarchy, on both sides of the Pacific. Of course, to have viewed these events chiefly through agents like Cameron Forbes is to engage empire through a distinctively skewed, top-down perspective. It has been the premise of the book, however, that the closely focused lens onto colonial power that such careers, cultural conventions, and thorough recordkeeping allow for, when adequately contextualized, has its virtues, providing an insider's view of American empire in the Philippines and, at last, onto the surprising precariousness of a governing regime. For his part, the once tireless road campaigner Forbes travelled halfway around the world but could not escape the precariousness of his own body.

Persistence of Empire

"I've felt miserable ever since I left," Forbes would confide in a June 1912 letter to Acting Governor-General Newton Gilbert during what was planned as a restorative visit, with brother Edward Waldo Forbes, to "the ranch" in Wyoming, "and in the depths of a gloomy depression, which is natural, I suppose."[24] The stimulation of high altitude failed to provide an anticipated environmental solution to Cameron's ongoing health struggle, and in late May he even regretted to Stimson that he was "still in very bad shape," and not up to providing a commentary on a Philippines-related House bill, as Stimson had requested. Although Forbes had updated Gilbert on June 15 that he was sleeping well and had regained his digestion, the next week's *New York Times* would report that Forbes had been confined on doctor's orders to his home on Commonwealth Avenue in Boston after "suffering from breakdown due to overwork in the islands, where he has been stationed for eight years."[25] Forbes's physician, his cousin Dr. Frank Watson, described the condition as "not serious," prescribing only undisturbed rest and relaxation at the sea shore. With Taft's assurances that he should take as long as needed to restore his health, Forbes would leave the everyday work of Insular administration in Gilbert's hands for several months longer while giving himself a chance to recuperate, under Watson's care, at the Forbes estate at Naushon Island, Massachusetts, taking few visitors and keeping business and rare trips to Boston to a minimum.

By the time Taft visited Forbes at Naushon in early October 1912—arriving on the presidential yacht *USS Mayflower* at nearby Woods Hole, after specifying that he did "not wish to undertake any horseback riding"[26]—many believed Taft had already given up on the bruising presidential campaign. Of course, neither Forbes's apparent breakdown nor the pending break-up of the Taft-Forbes regime signaled the end of *empire* in the Philippines. Indeed, if Forbes had been correct, in his 1909 inaugural speech, that capital required a stable government in Manila, then the U.S. presidential election might be understood as a "market correction." Perhaps the Democrats, the erstwhile anti-imperialist party, were best suited to running an "empire of no empire" after all, or to inching U.S. imperialism, in the Philippines and elsewhere, closer to a liberal American Empire, as Neil Smith posits, that merged ad hoc geopolitics with market-based power.[27] In the meantime, Dr. Watson's rest cure was apparently restorative. Forbes returned to Manila in January 1913 to resume the Governor-Generalship, taking the Pacific route, and by the end of the month could joke with cousin Nat Stone that he was back to "working eighteen hour days."[28]

Forbes had enlisted Stone in his plans for a retentionist, donor-supported "friends of the Islands" organization, possibly based on conversations with Taft, which featured a former *Manila Times* editor on retainer, to make a "determined campaign purely along the lines of education ..."[29] As Forbes intimated to Stone, "this is probably the crucial time in which the permanence of the work to which I have given myself entirely for nearly ten years now is to stand or fall, I am inclined to back this thing rather heavily."[30] Re-established at Malacañang, Forbes also set to work in efforts to install allies in the bureaucracy, including the Philippine Commission itself,[31] entrench signature policies in the landscape, and even to propose new ones. On February 23, Forbes pitched a vast and improbable scheme to Horace Higgins, General Manager for the British-owned Manila Railway Co., for a "Grand Central Terminal Company" set up to own: "1. A new water power to be developed; 2. The Street Railway of Manila; 3. The Electric Light of Manila; 4. All railroads in and about Manila; 5. The new piers; 6. The government arrastre or unloading plant at wharves; 7. A modern coal handling plant; 8. A set of modern warehouses for freight handling and storage."[32] While solving problems of labor and geography in one fell swoop through another reproduction of harbor space, the project, Forbes insisted, would make Manila "the most up to date and modern port in the Orient, and almost in the world"; construction costs would, admittedly, be high.[33] As late as August 1913, Forbes continued to push for an additional ten million pesos for the next Public Works budget, though by this time it was surely clear that his hopes of staying in office for a full year, serving as a bridging figure between administrations, would go unrealized.[34] An abrupt August 23 cable—inquiring about household linens, silver, and glassware at Malacañang, and requesting that Forbes engage servants for the incoming Governor-General before departing—reached Forbes *before* an August 25 letter (also cabled) from President Wilson thanking Forbes for his "faithful and careful service," and accepting his resignation effective September 1, provided the final insult. Forbes penned a 55-page letter on Insular policy, personnel, and the remodeling of Malacañang to Frank Carpenter, the Executive Secretary of the Philippine Commission, in case he might be "in a position to inform the new Governor-General of my ideas in regard to certain things in case he may ask you," and departed Manila in September in advance of the arrival of the monsoon and the new Governor-General, Francis Burton Harrison.[35]

Insular Regime Change

While the Philippines had not been a central issue of the 1912 presidential election, the Democratic Party had embraced the anti-imperialist planks of prior platforms, declaring itself "against a policy of imperialism and colonial exploitation in the Philippines or elsewhere. We condemn the experiment in imperialism as an inexcusable blunder, which has involved us in enormous expense, brought us weakness instead of strength, and laid our nation open to the charge of abandonment of the fundamental doctrine of self-government."[36] Whatever the constraints posed by this anti-imperialist stance, once in office Wilson's administration found itself in charge of an insular empire that included unincorporated territories under civilian (Insular) rule in Puerto Rico and under military rule in Guam, Samoa, the Panama Canal Zone, and Guantanamo Bay (in addition to incorporated territories of Hawaii and Alaska and "internal" sovereignty arrangements with Native American tribes). The presidency, in other words, *was* imperial. As the opening of the Canal in 1914 rendered the U.S. more functionally an inter-oceanic empire of trade networks and naval geopolitics, the Insular roster was expanded, before the end of Wilson's first term, initiating what would become lengthy military occupations of Haiti (1915) and Dominican Republic (1916), and the strong-armed, wartime purchase of the Danish West Indies (U.S. Virgin Islands) (1917).

Notwithstanding these emerging contradictions, the new Wilson-Harrison regime, in conversation with Manuel Quezon (as the Philippines' non-voting Resident Commissioner in the U.S. House of Representatives), sought to modify the nature of colonial governance in the Philippines: under Harrison, the Philippine Commission was reconstituted with a Filipino majority, offering Filipinos ostensible control of "both houses" for the first time. More controversial, in North America, was the Philippine Autonomy Act ("Jones Act"), passed in August 1916, which included, in a much-quoted preamble, a pledge of independence for the Philippines "as soon as a stable government can be established."[37] Perhaps more significant for Filipinos than vague promises, the Jones Act included substantive changes in the direction of greater autonomy at a national, archipelagic scale. The law, an organic act which replaced the 1902 Philippine Organic Act as the insular state's formal constitution, created an elected Philippine Senate, replacing the Philippine Commission (which would be dissolved) as the upper house of the Philippine legislature, while authorizing the Governor-General to appoint representatives from the "non-Christian" provinces.[38] Whereas previously the Army and Insular Government had maintained—and closely guarded—their territorial authority over the Special Provinces, henceforth the legislature's authority would extend over the whole archipelago.

After 12 years under Republican policy makers and appointees, it is hardly surprising that the new Insular Government would consider adjustments to the territorial arrangement of power in the Philippines, and through such changes, attempt to rework political, social, and economic relations in the archipelago in ways that might be squared with its own exceptionalist ideologies. If colonial empires persisted by producing—and reproducing—the spaces of their own survival, then the production of territory in the Philippines, as the space within which different forms of sovereignty could be exercised, reflected the legal basis of the colonial state itself, but it was subject to competing agencies and social relations both internal and external to state power. Territory, like other forms of state space, required upkeep and innovation amid conditions of uncertainty and change. The category of the *insular* was successful, in this sense, because of its versatility. Its value—as a naturalized, geographical name for civilian elements of the War Department—was as an "umbrella" category of bureaucracy that "covered" varied and dynamic local political arrangements; it named a flexible and evolving form of territoriality. In addition to changes in the territorial relations of democratic power under colonial sovereignty, the *new* Insular regime would seek to transform the embodiment of government labor through policies of accelerated "Filipinization" of the civil service which emphasized transition in leadership positions and a program of promotion and training geared toward more proximate temporal horizons of Philippine independence than the prior regime's typical projections, bracketed between a generation and a century. Beyond what was required by the Jones Act, Harrison's administration, as Forbes and General Leonard Wood, former military Governor-General of the southern archipelago (and earlier, Cuba), would later harangue, had "deliberately adopted the policy of getting rid of most of the Americans in the service, competent and otherwise," reducing the American colonial cadre from 28% to 4% of government workers by 1921.[39] This shift from "orderly promotion" to a "hurried Filipization," Wood and Forbes would insist, had injured the developmental legacy of the U.S. intervention in the Philippines, something they hoped to reverse in a Republican return to power in Washington.[40]

Extensive lobbying efforts against *Jones* by former colonial officials were driven, in part, by personal motivations, as leading colonial officials, deeply invested in the colonial project while fiercely defending their own records (and future career opportunities), rallied for retentionist colonialism with the support of the American Philippine business community and American Catholic church.[41] Taft himself, as early as November 1913, compromised norms of the "apolitical ex-presidency" in a widely-reported speech criticizing American-Philippine policy, later going so far as to personally lobby individual congressmen against the bill.[42] Dean Worcester, whose resignation, like Forbes's, was accepted in September 1913, had managed to acquire numerous properties in the Philippines (some under relatives' names) along with control of the newly capitalized American Philippine

Company, which would come to include a lucrative coconut oil refining central in Cebu, Mindanao cattle ranch and coconut plantations, and an inter-island shipping company that helped to vertically integrate his coconut empire.[43] Worcester also took to the lucrative North American lecture circuit, enticing audiences with new motion pictures—shot at government expense—coordinated with scripted lectures emphasizing Philippine wildlife and "wild peoples," and featuring bare-breasted women and girls, while arguing against the Jones Bill and Filipino capacities for self-rule.[44]

Perhaps the only thing worse, for the outspoken colonialists, than prematurely turning American colonial spaces over to the Filipinos had been turning them over to the new Insular regime—these spaces which, "during a moment of marvelous self-deception," they had claimed as their own, whether as aesthetic landscapes of sensory experience, mappings of the geo-body, or pathways of circulation.[45] And yet it was Harrison, a New York Congressman with family ties to Virginia planters, who would come to live in these inherited spaces as the longest serving American Governor-General, and to preside over the continuing development of Manila's waterfront esplanade, the clustering of scientific, medical, and university campuses around Ermita, and even, thanks to the Great War, the bolstering of Philippine exports in sugar, coconut oil, abacá fiber ("Manila hemp"), and cigars, along with the ensuing financial volatility that came with an economy pegged more closely to global trade. Or, rather, it was Harrison who let the Philippine legislature preside over these developments, from 1916 under the presidency of Quezon in the Philippine Senate and speakership of Osmeña in the House of Representatives; Harrison would exercise the gubernatorial veto just five times between 1913 and 1921.

Wood-Forbes Mission

If opposition to *Jones*, in the runup to the bill's passage in 1916, can be seen as a kind of ideological last gasp for advocates of long term, formal U.S. empire, that is not how colonial retentionists understood their situation at the time when, in the wake of Republican Warren Harding's landslide presidential election (and Republican retention of Congress), the "Grand Old Party" retook ownership of American Philippine policy after eight years in the political *bundok*. Perhaps based on Forbes's advice, President-elect Harding determined that a *study* of Philippine conditions would be the best means of swatting aside Wilson's lame duck December endorsement of Philippine independence.[46] After his inauguration in March 1921, Harding formally invited Forbes and Wood, a political competitor whom he had defeated in a brokered Republican Party convention, to lead the wide-ranging special investigation under a broad remit of reporting on conditions in the archipelago and offering recommendations. Its purpose was to advise the President whether, after eight years of Filipinization, the Insular Territory should indeed be deemed ready for the independence that it had indefinitely been promised.

Forbes's journals and correspondence suggest, however, that the fix was in from the start.[47] Interestingly, the invitation to join what would become known as the Wood-Forbes Mission would reach Forbes in the Caribbean, en route to New York from Venezuela ("a dream of tropical loveliness"), where he had been inspecting United Fruit Company properties, reflecting an apparent cross-over in methods and skill sets between the governmental and corporate worlds to which Forbes had returned.[48] The opportunity now to return to a regime of Philippine inspection in which not just the Filipinos but Harrison's entire succeeding administration could be placed under his scrutiny was far too good to pass up. After the War Department sorted out power sharing arrangements between the two principals,[49] Forbes was on board for a return trip to Manila (see Figure 5.1).

The Wood-Forbes Mission was inspection on steroids. Working out of Malacañang in Manila, the Mission conducted meetings and conferences with officials of the Central Government and "representative Americans, Filipinos, and foreigners of every walk of life," while devoting much of its

Figure 5.1 The Wood-Forbes Mission at Malacañang Palace, Manila (1921), in signed portraiture, featuring Leonard Wood (with riding crop) and W. Cameron Forbes (with cane) at center.

Source: Courtesy, Houghton Library (pfMS Am 2212 (74)), Harvard University.

time and efforts to on-site investigations in the provinces: a week in Manila, followed by two-to-four week excursions across the archipelago.[50] Over the course of four months from May to September, the Mission traversed some 15,000 miles by ship, train, automobile, horse pack, and dugout, the fate of a nation seemingly in tow (see Figure 5.2). Traveling both together and

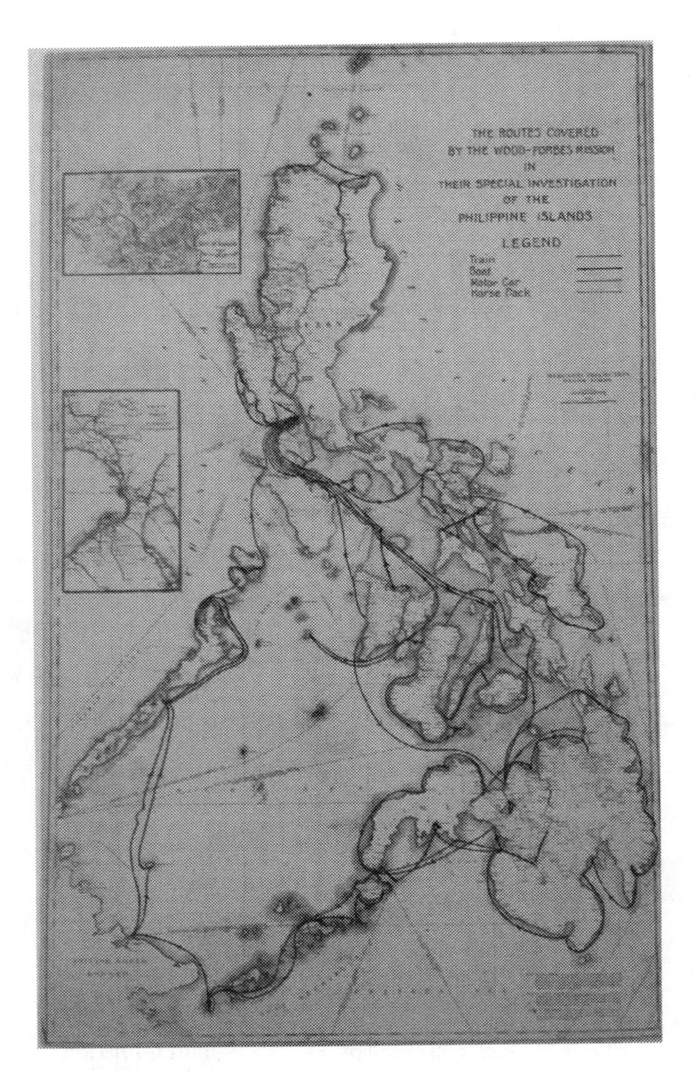

Figure 5.2 "The Routes Covered by the Wood-Forbes Mission" May to September 1921, with color-coded route arrows depicting modes of transport: primarily blue (boat) and red (motor car), with lesser arrows in green (train) and yellow (horse pack).

Source: Courtesy, Houghton Library (pfMS Am 2212 (83)), Harvard University.

separately, Wood and Forbes would visit an astonishing 449 municipalities in their investigations, including 46 out of 47 provinces plus Manila and Baguio, taking turns offering speeches in English and Spanish, and availing themselves for a torrent of "public sessions" and private interviews.[51] Yet despite overwhelming popular support for independence (with a military protectorate) from the Christian or "civilized" provinces, the Mission's report, in language that strongly evokes Forbes's prose, would respond to the question of independence with a finger-wagging "no," recommending instead that the "present general status of the Philippine Islands continue until the people have had time to absorb and thoroughly master the powers already in their hands."[52] Wood and Forbes, in fact, proposed a more reactionary turn, recommending that Congress declare null and void parts of *Jones* that diminished the authority of the Governor-General relative to the Philippine Legislature, but President Harding did not take up this agenda before his March 1923 death, in a San Francisco hotel, before what would have been the first presidential visit to Alaska Territory. The point of the Mission had been to hold the line, and *Jones* would continue to function in this sense as the constitution of the Philippines, an unincorporated Insular Territory, until the establishment of the Commonwealth in 1935. Immediately after the Mission's visit, Leonard Wood was appointed Governor-General.

On what basis had Wood and Forbes insisted that the U.S. must hold onto its colonial spaces in the Philippines? A premise of the investigation was that Philippine administration "in 1913, the period of greatest efficiency, was honest, highly efficient, and set a high standard of energy and morality." At the time, "Inherited tendencies were being largely replaced by American ideals and efficiency throughout the Philippine personnel, but the time and opportunity were both too short to develop experienced leaders and direction in the new English-speaking and American-thinking generation."[53] Even their instructions from Secretary of War John Wingate Weeks anticipated, under Filipinization, a "lowering of the standards of government" and "steady relaxation of effort," with "a more marked deficiency as we recede from the previous standard," as matters to be "carefully scrutinized."[54] The Mission's report offered the self-serving platitude that no people had ever advanced so quickly as the Filipinos under U.S. rule, and acknowledged successes of the government in the construction of new schools and extension of public education—which, at 945,000 enrolled, doubled 1913 levels—and measures of colonial political economy, including the accelerated growth of total business from $325 million in 1913 to $863 million in 1920, bolstered by sharp increases in U.S. trade.[55] But the failure of the Philippine National Bank—described in the *Report*, with excessive hyperbole, as "one of most unfortunate and darkest pages in Philippine history"—had provided the Mission with made-to-order proof of mismanagement (to the tune of perhaps $22.5 million in losses), was used to justify claims of pervasive "deterioration in the quality of public service," and "retrogression in the efficiency of most departments of the government" and the courts.[56] Blame, of course, did not rest "solely upon the Filipinos," for along with the "political infection" of the civil services under Filipino control—a critique that worked by placing the colonizer somehow outside of politics—the declining efficacy of Insular Government reflected the "bad example," "incompetent direction" and, above all, "lack of competent supervision and inspection," with Harrison singled out for ultimate responsibility.[57] Even the expansion to nearly 3,000 miles of first class roads, Forbes decried, had been enabled by the lowering of standards; indeed, during the Mission's excursions he had taken to chiding his hosts, as he had during his days as Secretary of Commerce and Police and Governor-General, whose roads he did not consider adequately maintained, or appeared only hastily maintained in advance of an inspection.[58]

"The Filipinos' Answer to the Wood-Forbes Report," a set of commentaries introduced to Congress by Resident Commissioner Jaime de Veyra, expressed disappointment in the Mission's recommendations, reiterating the Mission's own observations of the desire everywhere for independence among Christian Filipinos, even as "the pagans and non-Christians, constituting about 10% of the population," along with the "Americans in the islands, numbering 6,931 out of 10,956,730 total population … are for continued American control."[59] Meanwhile the erstwhile Forbes ally Osmeña, in a pointed critique of political agency expressed in the Wood-Forbes report, observed that, while credit was due to Forbes for his permanent road construction initiative, "it was the old Philippine Assembly which gave him the means to carry it out." Except for a few roads "like that of Baguio," Osmeña added, funds had not been systematically appropriated for public works before the Assembly in 1907, and since 1916, they had been extended in every Province.[60] More troubling, for Filipino critics, were the apparent efforts to "move the goalposts" on criteria for independence. As Maximo Kalaw put it in an extended commentary on the Wood-Forbes Report included in *The Filipinos' Answer*, "When President Wilson and Gov. Gen. Harrison officially certified that there was a *stable* government in the Philippines, they did not mean a *perfect* government."[61]

But if, as now seems certain, independence had never been properly on the table, then why the elaborate performance of the Wood-Forbes Mission? Why did it take the form that it did—why its particular geographical and epistemological trajectory? The Mission might have been understood predominantly as a conference or hearings-based inquiry located in Manila, for example, rather than a field-based investigation. Instead, "All parts of the archipelago were visited" so that Wood and Forbes could report to the War Department that: "your mission feels it has placed itself in intimate touch with the great mass of the Philippine people—Christian, Moro, and pagan—and with practically all Americans and foreigners domiciled and doing business in the principal cities and towns of the islands."[62] Even so, these geographically distributed connections with the Philippine masses would have been possible with the excursions limited to provincial capitals, or principal towns and cities. Visiting 449 towns and cities was an immense undertaking, carried out at a sometimes breathtaking pace of 10 stops per day, according to Forbes's journal.[63] For each site, the date of arrival was announced in advance to allow "ample time for the preparation of petitions, memorials, and addresses," which were presented in public or semi-public fora following speeches by Wood or Forbes (or both). "Almost without exception," the report describes, "the officials and people of the regions visited paid great attention to the reception of the mission. The roads and streets were decorated with arches, generally bearing the word, 'Welcome,' followed by a statement that the people desired their independence."[64] The visits frequently included site inspections of administrative and judicial offices, schools, hospitals, and jails, providing opportunities for firsthand observations of neglect, "retrogression," and decay. Private interviews were also held in most settings, and Forbes is at pains to document in his journal the sentiments conveyed by old acquaintances who privately assured him of their support for continued American rule, though such statements could not be said publicly. Or perhaps they were telling Forbes and the Mission what *they* wanted to hear, that is, playing to advantage their own positions in the new regime. Either way, it was precisely on the basis of these visits that the Mission had been able to "form definite conclusions" about conditions in the Philippines.[65] The first-hand observations in place, the traveling public and private spaces of political exchange, and ultimate exercise of an only ever *advisory* power—an authority of no authority—these events were precisely the point of the Wood-Forbes Mission and its broadly cast geographical expression. Reworking colonial practices of inspection in the context of Filipinization and independence debates, the Mission, though built around temporary appropriations of spaces and landscapes rather than more materially durable geographies of concrete and gravel, was another spatial strategy, a representation of democratic space produced—and circulated—to ensure the survival of U.S. empire in the Philippines, if only for another administration.

Of course, it was not all work and no play for the Missioners. Along with countless hours spent fishing and playing cards aboard the Polillo and other vessels, Forbes shot and collected 473 birds, comprising, he surmised, 220 varieties, some "potentially new to science," which he stored in a fine camphor box and crated off for Boston, "a job for our Harvard scientists."[66] Setting aside a likely concussion sustained at polo in Manila, Forbes judged the experience of the Mission "one of the wonderful experiences of my life," emphasizing camaraderie among "the men" while enjoying the venerations and gifts received from old acquaintances in Manila and the provinces.[67] The Mission did manage two visits to Baguio, where Forbes relished a "homecoming" with old friends. He was especially pleased to find Worcester "waiting at the top of the trail" as the party reached Baguio in late May over the Naguilian road and was struck by the new business magnate's apparent vigor, "prosperous, happier than ever before in his life; big, hearty, and full of cordiality and enthusiasm. We put in nearly the whole day together going over plans and events."[68] What a prospect it must have been. And yet, notwithstanding Worcester's success in anticipating markets in coconut oil and inter-island shipping, Philippine futures, and the future of U.S. empire in the Philippines, were uncertain, and an American imperialism in transition waited for no one. Dean Worcester would die unexpectedly at his home in Manila in 1924, following a short hospital stay, of chronic endocarditis and phlebitis at age 57. Leonard Wood, a contentious and unpopular Governor-General, died in office, on a brain surgeon's table in Boston, in 1927. Perhaps a bridging figure for liberal American empire after all, Cameron Forbes would go on to chair a 1930 U.S. commission investigating conditions in Haiti which advocated and planned for the withdrawal of U.S. Marines there, after nearly two decades of occupation, and to serve as Ambassador to Japan (1930–1932). He died in 1959, in his ninetieth turn around the sun, at his residence in the Hotel Vendome on Commonwealth Avenue.

***

Today a plaque dedicated to Forbes, containing at least one demonstrably false statement ("William Cameron Forbes was governor-general of the Philippines from 1900-1913"), is affixed to a large gray rock on the premises of the Good Shepherd Convent, encompassing Forbes's former Topside retreat in Baguio. The large stone bungalow, if it is still standing, is difficult to discern among the Convent's spaces, which include the Good Shepherd Sisters' Mountain Maid Training Center, famed for the seasonal jams and medicinal teas which they have sold on site since the 1950s. It is fair to presume, however, that the extensive view of north-south ridges of the Cordillera Central is much the same as that seen from Topside's breakfast porch during Forbes's day, even if the illusion of ownership of this spectacular landscape is more difficult to maintain. While Burnham's initial town plan for Baguio had anticipated a population "not exceeding 25,000 inhabitants,"[69] Baguio City is now home to 366,000, and like many of the valley walls surrounding the old town, multi-story homes, apartments, and hotels have been densely constructed on the steep hillsides around Topside's perch (see Figure 5.3). Rather than a sense of ownership, gazing over this landscape evokes an aesthetic of accretion.[70] And while it may be true that all that is solid in spaces and landscapes, to paraphrase Marx, someday melts into air, elements endowed with differential material and symbolic durability and heft, more specifically, remain to be reworked, re-appropriated, struggled over, or reproduced, including history itself. Whatever the errors of history on the Forbes plaque, however, the strawberry jam and turmeric tea and from Good Shepherd Convent were delightful and the view was magnificent.

In *American Colonial Spaces in the Philippines*, I have attempted to tell a series of plausible stories, constructed from the historical record, about a

Figure 5.3 View from the Good Shepherd Convent, Baguio, near the site of the former *Topside* residence.

Source: Photo by author.

set of spaces that no longer exist. Though some of their traces and impacts remain, the point of the book has not been an impact assessment. Rather, the *making* of territory, maps, landscapes, and roads, and how such spaces were understood as solutions to problems of colonial rule, has been the point, the story of stories that I have attempted to narrate. The analysis presented has largely been limited to a particular "imperial moment," but the questions raised—about how the production of different kinds of space makes possible the reproduction of relations of power over time—are by no means foreclosed in the shift to different forms of empire, imperialism, and the state, even as "American" colonial spaces were reproduced as new state spaces under the Philippine Commonwealth and Republic. What is more, the two countries have remained deeply entangled in the production of spaces that enabled the reproduction of social, political, and geopolitical relations in the Philippines (and Southeast Asia), from independence in 1946 to martial law in 1972 to the expulsion of U.S. forces from the Subic Bay and Clark military bases in 1991 and beyond.[71] The United States, though it may at times seem invisible, also *remains* an insular empire, characterized by unequal political relations, rights, and protections in the insular "possessions," in a changing world wherein tropical islands have continued to hold special significance as means of projecting geo-strategic power.[72] Though historically shifting, these "geographical crumbs," as Smith described, more specifically, the territorial spoils of the Spanish-American War,[73] are in fact among the "American century's" enduring spatial elements, a geographical infrastructure of formal empire that makes the larger "empire of no empire" possible.

Notes

1 Gilman to Heiser, April 19, 1917. W. Cameron Forbes Papers, bMS Am 1364.4, Box 4, Houghton Library, Harvard University.
2 Forbes to Edwards, November 23, 1911. W. Cameron Forbes Papers, MS Am 1366.1, Confidential Letter Book No. 1, Houghton Library, Harvard University.
3 William Cameron Forbes, "Journal," (entry 12/10/1911). First Series, Vol. V, p. 77. W. Cameron Forbes Papers (1930), MS Am 1365, Houghton Library, Harvard University.
4 Ibid.
5 Ibid.
6 Forbes to Stimson, December 11, 1911. W. Cameron Forbes Papers, MS Am 1366.1, Confidential Letter Book No. 1, Houghton Library, Harvard University.
7 Ibid.
8 Ibid.
9 William Cameron Forbes, "Journal." First Series, Vol. V, pp. 126–155. W. Cameron Forbes Papers (1930), MS Am 1365, Houghton Library, Harvard University. Philippine Commissioner Newton Gilbert served as Acting Governor-General in Forbes's absence.
10 William Cameron Forbes, "Journal," (entry 3/22/1912). First Series, Vol. V, p. 125. W. Cameron Forbes Papers (1930), MS Am 1365, Houghton Library, Harvard University.
11 Forbes, "Journal," (entry 3/22/1912), Vol. V, p. 125.
12 Forbes, "Journal," (entry 3/22/1912), Vol. V, p. 126.
13 Ibid.
14 Forbes, "Journal," (entry 3/22/1912), Vol. V, p. 127.
15 Forbes, "Journal," Vol. V, p. 131.
16 Forbes, "Journal," Vol. V, p. 135; Forbes to Stimson, June 17, 1912. W. Cameron Forbes Papers, MS Am 1366.1, Confidential Letter Book No. 1, Houghton Library, Harvard University.
17 Forbes, "Journal," Vol. V, p. 145. "Westward the course of empire takes its way" references the 1860 mural by Emanuel Gottlieb Leutze, an emblem of American 'manifest destiny' in North America.
18 Forbes, "Journal," Vol. V, p. 144.
19 Forbes, "Journal," Vol. V, p. 145.
20 Forbes, "Journal," Vol. V, p. 147.
21 Forbes, "Journal," Vol. V, p. 154.
22 Forbes describes Taft as "fearfully preoccupied with the disaster that had overtaken him in having his … chief turn against him" and disinclined to "come to matters of business." Forbes, "Journal," Vol. V, p. 155. Theodore Roosevelt's public criticism of Taft, his onetime protégé, presaged his walking out of the Republican convention the next month to launch his Progressive, third-party challenge for the presidency.
23 Henri Lefebvre, *State, Space, World: Selected Essays*, eds. N. Brenner and S. Elden (Minneapolis, MN: University of Minnesota Press, 2009), pp. 95–123, 138–152.

24 Forbes to Gilbert, June 15, 1912. W. Cameron Forbes Papers, MS Am 1366.1, Confidential Letter Book No. 1, Houghton Library, Harvard University.
25 Forbes to Gilbert, June 15, 1912; "Gov. Forbes Ill: Worn Out by Eight Years Under High Tension in the Philippines" *New York Times*, June 22, 1912, p. 7.
26 Forster to Taft, [October] 11, 1912. W. Cameron Forbes Papers, Letters—Prominent (1364), File 304, Box 7. Houghton Library, Harvard University.
27 Neil Smith, *American Empire: Roosevelt's Geographer and the Prelude to Globalization* (Berkeley, CA: University of California Press, 2003).
28 Forbes to Stone, January 26, 1913. W. Cameron Forbes Papers, MS Am 1366.1, Confidential Letter Book No. 2, Houghton Library, Harvard University.
29 Forbes to Higginson, January 26, 1913. W. Cameron Forbes Papers, MS Am 1366.1, Confidential Letter Book No. 2, Houghton Library, Harvard University. "Friends of the Philippines" fits with the agenda which Forbes had described, in a December 1912 with Taft, as "the campaign for the Philippine Islands." Forbes to Taft, December 9, 1912. W. Cameron Forbes Papers, MS Am 1366.1, Confidential Letter Book No. 1, Houghton Library, Harvard University. On Taft's activist post-presidency and the retentionist movement, see Adam D. Burns, "Retentionist in Chief: William Howard Taft and the Question of Philippine Independence" *Philippine Studies: Historical & Ethnographic Viewpoints* 61 (2013): 163–192.
30 Forbes to Stone, January 26, 1913.
31 These efforts, including attempts to appoint a crony (and former personal secretary), Edward "Pete" Bowditch, Jr., as the first civil Governor of Moro Province, began before Forbes left American shores. Forbes to Taft, December 9, 1912.
32 Forbes, "Journal," (entry 2/23/1913), Vol. V, p. 191.
33 No response from Higgins, who was "sending home recommendations on the matter," is indicated. Approval from legislators for the vast scheme to rework both the physical environment and social relations of labor on the docks, as Forbes later acknowledged, was unlikely. Forbes, "Journal," (entry 2/23/1913), Vol. V, p. 191.
34 Forbes to Egan, August 8, 1913, W. Cameron Forbes Papers, MS Am 1366.1, Confidential Letter Book No. 2, Houghton Library, Harvard University.
35 McIntyre to Forbes, August 23, 1913, and Wilson to Forbes, August 25, 1913. W. Cameron Forbes Papers, MS Am 1364, Box 8, Houghton Library, Harvard University; Forbes to Carpenter, August 25, 1913, W. Cameron Forbes Papers, MS Am 1366.1, Confidential Letter Book No. 2, Houghton Library, Harvard University. Carpenter was appointed the first civilian Governor-General of Moro Province in December 1913.
36 1912 Democratic Party Platform, June 25, 1912. Available via The American Presidency Project, UC Santa Barbara: https://www.presidency.ucsb.edu/documents/1912-democratic-party-platform.
37 U.S. Congress, *An Act to Declare the Purpose of the People of the United States as to the Future Political Status of the People of the Philippine Islands, and to Province a More Autonomous Government for those Islands*, Public Law No 240, August 29, 1916.
38 In the shift to a bicameral legislature, the Philippine Assembly was renamed the House of Representatives.
39 *Report of the Special Mission on Investigation to the Philippine Islands to the Secretary of War* (Washington, DC: U.S. Government Printing Office, 1921), pp. 16–17.
40 Ibid., p. 17.
41 Burns, "Retentionist in Chief"; see also Paul A. Kramer, *The Blood of Government: Race, Empire, the United States, & the Philippines* (Chapel Hill, NC: University of North Carolina Press, 2006), pp. 357–388.

42 Burns, "Retentionist in Chief."
43 Rodney J. Sullivan, *Exemplar of Americanism: The Philippine Career of Dean C. Worcester*, Michigan Papers on South and Southeast Asia 36 (Ann Arbor, MI: Center for South and Southeast Asian Studies, University of Michigan, 1991), pp. 191–213.
44 Pan American Film Co., "Native Life in the Philippines" (advertisement for film and lecture), Box 3, Dean Conant Worcester Papers, Bentley Historical Library, University of Michigan; Sullivan, *Exemplar of Americanism*, pp. 165–190; Mark Rice, *Dean Worcester's Fantasy Islands: Photography, Film, and the Colonial Philippines* (Ann Arbor, MI: University of Michigan Press, 2014), pp. 118–155.
45 Henri Lefebvre, *The Production of Space*, trans. Donald Nicholson-Smith (Oxford: Blackwell, 1991), p. 189.
46 Forbes to Harding, February 7, 1921. In Forbes, "Journal," ("Wood-Forbes Mission – Appendices"), Second Series, Vol. II, pp. 313–315. W. Cameron Forbes Papers, MS Am 1365, Houghton Library, Harvard University.
47 Indications that a rejection of Wilson's proposition (and Philippine independence) was a foregone conclusion include a letter, prior to the journey, from former Philippine Governor-General Luke Wright, concurring with Forbes that: "The view that you express that it is a piece of criminal folly to grant independence now seems to me incontestable." Wright to Forbes, February 9, 1921. In Forbes, "Journal," Second Series, Vol. II, p. 37. Forbes expressed comfort with how much the he and Wood were "in accord. I found our ideas tallied very closely as to fundamental conditions so far as our job and the remedies to the present situation were concerned." Forbes, "Journal," Second Series, Vol. II, p. 43.
48 Forbes, "Journal," Second Series, Vol. II, p. 23.
49 Wood was designated Chairman of the Mission, but not its Chief, with the two awarded equal votes, and equivalent diplomatic ranks of Governor-General, in reporting and signing onto the Mission's recommendations. Forbes, "Journal," Second Series, Vol. II, p. 42.
50 *Report of the Special Mission*, p. 11.
51 Ibid.
52 Ibid., p. 46.
53 Ibid., p. 22
54 Ibid., p. 9.
55 Ibid., pp. 17, 33–34. Philippine trade surpassed $300,000,000 in 1920, roughly two-thirds of which was with the United States.
56 Ibid., pp. 38, 17, 20.
57 Ibid., pp. 20, 22, 45–46.
58 Ibid.; Forbes, "Journal," Second Series, Vol. II, p. 102, 118.
59 de Veyra and Gabaldon to Harding, December 13, 1921. In *The Filipinos' Answer to the Wood-Forbes Report: Remarks of Hon Jaime C. de Veyra of the Philippine Islands in the House of Representatives,* January 5, 1922 (Washington, DC: U.S. Government Printing Office, 1922), p. 4.
60 In response to the report's sharp critique of the Philippine judiciary, the erstwhile Forbes ally Osmeña also noted, in a December 1921 address in the wake of the report's release that was summarized in the *Manila Times* (December 4, 1921), that "anybody attacking our courts must appeal to the proper authorities and produce facts for the removal of incompetent and immoral judges, and should he fail to substantiate his charges he alone must be held responsible for the consequences of his act." In *The Filipinos' Answer*, p. 6.
61 Emphasis added. Maximo M. Kalaw, "The Wood-Forbes Report—A Critical Analysis" in *The Filipinos' Answer*, p. 7. Kalaw was Dean of the University of the Philippines' College of Liberal Arts.
62 *Report of the Special Mission*, p. 11.

63 Forbes, "Journal," (entry 7/06/1921), Second Series, Vol. II, p. 84.
64 *Report of the Special Mission*, p. 12.
65 Ibid.
66 Forbes, "Journal," Second Series, Vol. II, p. 168.
67 Ibid., p. 158. Gifts included a walking cane with head of Benguet gold from his "Igorot friend Jose Pianaza Kuknong" and numerous small presents, sometimes offered by Filipina 'granddaughters' whom elite families were delighted to show off to 'Papa Forbes.' Ibid., p. 168.
68 Ibid., p. 81.
69 Daniel H. Burnham and Pierce Anderson, "Preliminary Plan of Baguio Province of Benguet P. I.," June 27, 1905, p. 1. Daniel H. Burnham Collection, Box 56, FF 3. Ryerson and Burnham Archives, Art Institute of Chicago.
70 See David Lowenthal, *The Past is a Foreign Country—Revisited* (Cambridge: Cambridge University Press, 2015), pp. 122–126.
71 As reflected in recent work by Colleen Woods, *Freedom Incorporated: Anticommunism and Philippine Independence in the Age of Decolonization* (Ithaca, NY: Cornell University Press, 2020); and Mike B. Hawkins, "From Colonial Cargo to Global Containers: An Episodic Historical Geography of Manila's Waterfront," unpublished PhD dissertation, Department of Geography, University of North Carolina at Chapel Hill (2022).
72 Administered, paradoxically, under of the U.S. Department of Interior, the Office of Insular Affairs (the post-World War II successor to the Bureau) continues to coordinate U.S. federal policy in the territories of American Samoa, Guam, U.S. Virgin Islands, and the Commonwealths of Puerto Rico and the Northern Mariana Islands, and to administer U.S. federal programs in the Republics of the Marshall Islands, Micronesia, and Palau, under varied sovereignty agreements. For a recent assessment, see Sasha Davis, *Islands and Oceans: Reimagining Sovereignty and Social Change* (Athens, GA: University of Georgia Press, 2020).
73 Smith, *American Empire*, p. 16.

Index

Note: *Italicised* folios refers figures and with "n" refers notes.

Printed in the United States
by Baker & Taylor Publisher Services

Index

P

R

S

T

W

Z